AMERICA:
THE
OIL HOSTAGE

*From Oil Hostage to Oil Freedom
in a Generation.*

"America: The Oil Hostage," by Paul Bures. ISBN 1-58939-945-5 (softcover); 1-58939-946-3 (hardcover).

Library of Congress Control Number: 2006935620

Published 2006 by Virtualbookworm.com Publishing Inc., P.O. Box 9949, College Station, TX 77842, US. ©2006, Paul Bures. All rights reserved. No part of this publication may be reproduced, stored in a retrieval system, or transmitted in any form or by any means, electronic, mechanical, recording or otherwise, without the prior written permission of Paul Bures.

Manufactured in the United States of America.

AMERICA: THE OIL HOSTAGE

From Oil Hostage to Oil Freedom in a Generation

by Paul Bures

*To My Loving Wife,
Sherri Lynn*

Table of Contents

Introduction 1
1. Oil Basics 9
  Origins of Fossil Fuels – the Sun's Buried Energy . 9
    Petroleum 11
    Oil History 14
    Oil Production 17
  Refinery 19
  How much Oil is in the United States? 21
  United States and Unconventional Reserves 23
    Oil left in the Ground 24
    Oil Sands of Athabasca, Canada 24
    United States Oil Shale 26
    Oil Supply vs. Oil Demand 28
    Other Oil Theories 31
2. Oil War Games 37
  Cost of Oil Imports 37
    Cost of Terrorism 38
    Cost of Blackmail 43
    The Cost of War and its Consequences 44
    The Cost of Government Friction 46
    Economic Instability 48
    Future Oil Prices 51
    National Security 53
    Rise of OPEC 55
    Free Trade vs. Protectionism 60
    Free Trade 61

Protectionism 63
Oil - a Commodity 65
Perfect World and Free Trade 68
U.S. Oil Companies – OPEC Connection 71

3. Caught Off Guard **75**
President Eisenhower 78
President Richard Nixon/President Gerald Ford .. 78
President Jimmy Carter 79
President Ronald Reagan 82
President George Bush Sr. 83
President Bill Clinton 85
President George W. Bush 86

4. Other Energy Sources **106**
Natural Gas 107
Coal 108
Nuclear Energy 110
Wind Energy 114
Solar Energy 118
Biomass 121
Hydropower 124
Geothermal Energy 128
Energy Use Projections 130

5. Hydrogen Solution (The Iceland Way) **132**
Hydrogen vs. Electric 132
Hydrogen Fuel 134
Fuel Cell 139
Hydrogen Economy (The Iceland Way) 141

6. Electric Solution (The Brazil Way) **145**
Introduction to Electricity 145
Electric Availability 147
Fossil Fuel Power Plant 149
Problems with Electricity 150
Batteries 152
Battery Electric Vehicle (BEV) 156
Hybrid Electric Vehicle (HEV) 159
Plug-in Hybrid Electric Vehicle (PHEV) 160
Electric Solution (The Brazil Way) 162

Introduction   5

7. Conservation Solution (The U.S. Way)......... 171
  CAFÉ (The Corporate Average Fuel Economy
  Program).. 172
  Increase Fuel Efficiency by Modifying CAFÉ...... 181
  License Plates Solution 185
  Award Conservation Solution........................... 189
  Conservation Using Batteries........................... 192
  Conservation Conclusion 197
8. When All Else Fails (The Europe Way) 199
  10% Solution.. 203
  Side by Side with Europe 208
9. Fun with Numbers 212
  Alternative Solutions 212
    Electric Power Plant Production 214
    Wind Energy 215
    Solar Power.. 215
    Biofuels or Ethanol.. 216
  Alternatives Combined..................................... 217
    Increase the Oil Production 217
    Conservation Control Commitment 220
    Electric Production 222
    Electric Surplus... 222
    New Power Plants.. 223
    Wind.. 223
    Solar.. 224
  More on Conservation and Efficiency 225
Conclusion 228
Index.. 240

Introduction

"**A**merica: The Oil Hostage." What is this all about? The strongest country in the world, the most technologically advanced country in the world, and the most admired country in the world – a hostage? That does not sound too thrilling but it is exactly what is happening. Most Americans do not realize that the United States imports two-thirds of the oil they use from other countries. Every time they fill up their gasoline tanks, two of every three gallons came from somewhere else in the world. Hence, when some unrest brews anywhere in the world, no matter how small, the U.S. oil supply is threatened.

These supplies are not under direct control of the U.S. government or the U.S. oil companies. Because these countries control the oil supply, they also control the price, and the U.S. government and the oil companies have to negotiate for both. The challenge is with the countries that the U.S. has to deal with, because many of these countries are just plain unfriendly to the United States. Some of these countries are outright hostile but as long as the United States continues to need their oil the U.S. will remain at their mercy. Thus, "The Oil Hostage."

With all the attention oil is getting these days, one would think that the U.S. government would want to come to America's rescue and solve the energy situation. All of the attention given to the subject of oil by the U.S. government and the media suggests that something is actually being done. Unfortunately it is all just an illusion. The fact is that this government tries to avoid the oil problems like an epidemic. The U.S. government knows the oil days are numbered and soon oil alternatives will need to be found, however they do not know of any quick solutions. They know that solving America's oil problem will take a long time and will require resolve and cooperation of all branches of the government, including the resolve and cooperation of both political parties. They also recognize that true solutions will be unpopular, therefore no one from either political party will be willing to be first, because if they do, "they are out!" And you know what that means.

So instead, Americans are being offered temporary quick fixes. When the price of oil spikes, the government usually offers a variety of different solutions, which look as though they were designed to curtail the price of oil and solve the oil dilemma. Every year the lawmakers return to the same challenge of seeming oil crisis, introducing different ideas. In 2004, the big push was to release some oil from the U.S. oil reserves, temporarily increasing the oil supply to lower the price. In 2005, the push was to establish a windfall tax on oil companies and to distribute the collected funds to Americans.

And in 2006, an election year, the individual political parties were promising to cure all of the oil ailments if reelected. The democrats proposed to temporarily eliminate the fifty-cent gasoline tax that Americans currently pay on every gallon of gasoline, and the republicans, to trump the democratic scheme, offered to send a rebate check between $100 and $300 to all American drivers. Neither of these had any chance of passage. The current buzz coming from the policy makers in Washington is ethanol, which is also unlikely to contribute to any meaningful oil solution (see Chapter 4, Ethanol). Despite that, however, many politicians will find a place for ethanol in their reelection platforms for years to come.

Ethanol is easy to understand and even easier to explain, so do not be surprised when even the presidential candidates begin to mention ethanol as a solution to the upcoming energy crisis. The candidates of both political parties will likely promote ethanol as a solution to the upcoming oil shortages and rising oil prices, never blinking an eye that they are actually misleading the public into a false sense of security as it pertains to the world's availability of oil. And so, even as the U.S. faces the possibility of oil prices skyrocketing, fivefold or even higher, the political friction continues and the oil dilemma remains unresolved.

The business community and the media are echoing the government's inactions. Every time the price of oil reaches a new high they all act stunned. When the price of oil reached $50 in 2004, everyone in the media dismissed the occurrence as a fluke, predicting the price would drop back down. When the price of oil reached $70 in 2005, some began to echo the possibility of an economic depression. And in 2006, with the price of oil in the high seventies and on the way to likely surpass $100 in the next couple of years, they all act dazed and astonished yet again.

For now it is just a game. The fact is that everyone expects new highs of crude oil to occur on a regular basis, at least every time a new summer driving season approaches. The government and the oil community know the prices of oil will continue to increase regularly in the future. They know that one day the quick fix will not be enough and the oil problem will have to be looked at seriously. Eventually the game of small oil fixes will turn into a serious problem where everyone will need to get involved to produce a solution. The question is; what price of oil will trigger this to happen?

The fact is that higher oil prices initiate discussions of alternative fuels. Therefore, should the price of oil temporarily retreat from the high seventies to the low sixties, bringing the price of gasoline back down to around $2 per gallon, all of the talk will end and any progress that would get the U.S. closer to oil independence will disappear until the next time around. It could take up to twenty years to replace the imported oil with domestically produced alternative fuel, so when the U.S. lawmakers finally get serious due to the onset of painfully high oil prices, the American people will continue to suffer for a long time. It sure seems that the path to oil independence should begin now, while the price of oil is fairly low.

Introduction

Is it too late? Is America at a point where there is no solution that can prevent a disaster? No it is not, it just may not be as easy an undertaking as it would have been had the U.S. started thirty years ago when the problem was first realized. By starting thirty years too late, Americans may have to pay a price and may have to sacrifice a bit. The price Americans will have to pay today will be dwarfed when compared to the price they will have to pay if the solution to the problem is delayed any further.

Having started thirty years too late, the U.S. has to continue to guard their oil supply, thus the oil wars of today. These oil wars are the unfortunate consequence of oil dependence and will get worse before the situation gets better. Many Americans have died in these oil wars and many more will die by the time it is all over.

Oil independence, which would help lead to oil freedom, is the only solution. The bright side of it all is that once the solution, a path to oil independence, is adopted and begins to be put into action, the oil wars can end. Once the implementation phase begins there will be no need to continue to secure more oil for the ever-increasing U.S. demand. The flip side, however, is that until that happens the vicious and deadly wars will continue and soldiers will keep on dying.

There are many solutions available to the U.S. government to bring the United States back on a path to oil independence to gain oil freedom. That is the main purpose of this book. None of these solutions are new or newly invented; all of them have already made their debut, in some form or another, somewhere in the U.S. economy or in different countries of the world. These fixes just expand on the successes of these programs and adapt them to the current oil situation in the U.S. This book will provide a comprehensive overview of the current oil situation in the United States. It includes just enough details to reach thorough understanding, in simple non-scientific language, about oil and all of the available oil substitutes that exist today. It will touch on all of the oil options; the common sense solutions of today and the futuristic solutions of tomorrow. Reading it will enable you to recognize, when listening to the energy discussions on TV, who is addressing the oil situation seriously and who is not.

The first chapter presents oil basics. It discusses where oil comes from, how it is made and how much of it is left. It explains who owns the oil and how is it currently being sold on the world's open markets. Chapter two introduces and argues the case of why oil should be replaced with alternative fuel and why the U.S. should become oil independent. Chapter three reviews what has been done to date; what the government and the oil companies have done in the past thirty years. Chapter four summarizes all other energy sources currently available, which have the ability to replace oil. Chapter five explains the faults and benefits of the hydrogen economy and the probability that it could be the future of the United States.

Chapters six through eight will offer simple, easy-to-follow solutions to gaining oil independence. They will show how simple the process of eliminating the need for imported oil could be if the government was to take the initiative and pass a few simple laws. Chapter six (The Brazil Way) introduces the electric economy. It will discuss it from the standpoint of having an automobile of the future capable of running on both gasoline and electric power. Chapter seven (The U.S. Way) is all about conservation. It reviews what conservation steps have been taken in the past and how successful they were. It will introduce various conservation proposals to oil independence and will offer a few specific examples. Chapter eight (The Europe Way) is about how when all else fails, you should increase the price of oil the way Europe did. Chapter nine is just fun with numbers, where actual examples are used to put all the loose pieces together.

The conclusion will introduce and describe America's bright energy future. In addition to a four-step plan to reach oil independence, it will summarize the numerous sources of energy that can replace oil and the rest of the fossil fuels. It shows that not only can these sources of energy replace oil, which is running out, but also that each source individually could replace oil by providing a clean, pollution-free, unlimited source of energy. It will point out that even though the U.S. could pick just one of these sources to satisfy America's energy needs, utilizing all of the newly described energy sources simultaneously can provide America with enough clean energy to last forever. In fact, utilizing these sources could provide enough energy to satisfy the needs of the entire planet in unlimited quantities lasting indefinitely.

With the availability of the Internet, there exists a vast resource of information for research. In this book only the basics on all of the subjects are discussed. However, if anyone is interested in additional more specific scientific and technical information on various subjects that have been introduced, various links have been provided at the end of each chapter, all of which were researched and consulted for information in this book. Nuclear energy, for example, was given attention on only four pages, yet the links provided introduce hundreds of additional pages filled with detailed technical information, which is beyond the scope here.

"Wikipedia, the free encyclopedia," has been the source of information on many subjects. It was an invaluable source in that it not only provided information on specific subjects, but it provided numerous additional links on a variety of related topics. It even provided a list of various books available focusing on individual issues. The web pages of the Energy Information Administration (EIA) and the U.S. Department of Energy (DOE), on the other hand, contain thousands, perhaps tens of thousands, of pages of data, tables, graphs, articles, and reports of all of the energy sources. In addition to current data, these sites provide historical data going back several decades. They provide data of projected future use and consumption of all energy used, no matter how minor a part they may have in the overall energy picture. Most of the specific data on production, consumption, and projected use of energy, has been obtained from these sites. You may be well-advised to visit these pages and perhaps get lost in the sea of information. So let's begin and see what the stuff that made America "The Oil Hostage" is all about.

1.
Oil Basics

Origins of Fossil Fuels – the Sun's Buried Energy

Astronomers estimate the earth to be about 4.5 billion years old. Up until 500 million years ago, algae, seaweeds, sponges and jellyfish were the only living things in our oceans. Land life began around 500 million years ago, when insects, animals and plant life began to evolve from sea algae. Earth's dinosaur era began 230 million years ago and lasted over 160 million years. Finally, with the extinction of dinosaurs some 65 million years ago, the new era that followed was the rise of modern birds, mammals and humans.

None of this could have been possible without the sun's energy. The sun is the source of earth's energy. The sun's energy hits the earth in the form of electromagnetic radiation we call sunshine, which plants and animals need to live and grow. Plants have developed a way to take the sun's energy, along with molecules from air and water, and turn it into energy that animals and humans can use. This process is called photosynthesis.

Photosynthesis provides a way to turn the sun's energy into a usable form of energy that not just plants but also the rest of the animals, including humans, can use. Photosynthesis begins when sunlight hits a plant's leaves. Chloroplast within the plant's leaf takes in water (H_2O) through the plant's root system and carbon dioxide (CO_2) from the air, and uses chlorophyll to absorb sunlight energy. With the help of sunlight energy, chloroplast re-arranges the water molecule and carbon dioxide to produce a carbohydrate molecule (CH_2O). Carbohydrate molecules combine to form glucose ($C_6H_{12}O_6$), a simple sugar. Finally, the plant releases the unused oxygen (O_2) left over from carbon dioxide back into the air. The released oxygen, which is essential to life on this planet, essential to all living and breathing things, replenishes the oxygen in the earth's atmosphere.

At the end of the process, the sugar molecules in a sense store the sun's energy in the form of chemical energy, or potential energy that can be used later. Plants use some of this energy for themselves and store the rest in the form of roots, fruit and seeds. Fortunately for life on earth, plants produce more food then they can use themselves, therefore creating an overabundance of food for everyone.

When animals feed on plants, they are in fact transferring the sun's energy to their bodies. They use up this stored energy, through the process of metabolism, for their nourishment to grow and survive. Metabolism and the process of glycolysis is essentially photosynthesis in reverse. Metabolism breaks down the sugar molecule to form water and carbon. The carbon is then exhaled as carbon dioxide.

To summarize, carbon dioxide, which was taken from the atmosphere, was needed to provide carbon to create sugars that store the sun's energy. Later, once the sun's energy is used up, it is passed on through the process of metabolism to animals, in the form of nourishment. The carbon is then released back into the atmosphere as carbon dioxide, thus completing the "carbon cycle." Plant life, however, continues to hold onto the sun's energy even after it dies, making the formation of petroleum possible.

Petroleum

All fossil fuels have been formed from dead plants and animals. They fall into three categories. 1) Petroleum, which includes liquid oil, gaseous natural gas, and bitumen, a tar-like solid that has been formed in ancient seas through the decay of plants and algae. 2) Coal, a solid, has been formed in land swamps through decay of forest trees and ferns. 3) Oil shale, a solid rock formed on land through decay of algae but lacking the heat and pressure that helps create oil, must still go through the process of distillation before it can be used. This section deals primarily with petroleum; the other natural resources will be discussed later in chapter four.

Oil was formed from remains of plants, algae, and prehistoric marine animals that lived in lakes and ancient seas millions of years ago. These remains settled to the bottom where they were buried in sand and mud and covered by layers of other sediments. Rivers have very slowly added many types of deposits as they washed off mud and sand to the seas and oceans. Over time, during a very lengthy process, these remains decayed and formed a layer of fine-grained shale mixed with organic matter. As millions of years passed, many additional sediment layers formed above, causing pressure and heat to increase significantly. This increased pressure and heat transformed this organic matter, trapped in rock, to kerogen (organic matter within a sedimentary rock). This process is known as catagenesis. In a final step, kerogen transforms into liquid and gaseous hydrocarbons.

Large pressures from above forced these hydrocarbons out of the source rock to the surface through the porous layers of sandstone and limestone. Most of these hydrocarbon deposits escape to the surface where they dissipate, however, when certain conditions develop, some of these hydrocarbons will accumulate and form underground reservoirs. These underground reservoirs form when oil and gas seeping to the surface meets an impervious, or non-porous, rock and becomes trapped. These traps were formed due to massive movements of the earth's crust, in which layers of rock slid past each other, forming various types of impenetrable traps. These are the very traps the oil companies search for, so they can recover the natural gas and oil within.

Different pressures and heat transformed this organic material into different forms of hydrocarbons. Today's hydrocarbons, the non-renewable natural resources, are found as natural gas, crude oil and bitumen. When these compounds are burned, the process transfers the elements within to its original state, where the carbon trapped within is released, thus completing the "carbon cycle" once again. In the burning process, hydrogen combines with oxygen to form water and carbon combines with oxygen to form carbon dioxide, creating heat that is collected and used. Burning fossil fuels is inherently recreating the atmosphere as it existed millions of years ago.

Burning hydrocarbons produces carbon dioxide (CO_2), which is one of the greenhouse gases contributing to "global warming." The increased concentration of carbon dioxide in the earth's atmosphere absorbs any escaping infrared radiation, enhancing the green house effect. This causes the earth's atmosphere to get warmer. In the last 150 years, the concentration of carbon dioxide in the earth's atmospheres increased by 40 percent. Fossil fuels are responsible for 7,000 million metric tons of CO_2 per year, and while the oceans will eventually absorb most of the carbon dioxide, it will take in excess of one hundred years. Sulphur dioxide is another major pollutant, created by burning hydrocarbons. This pollutant dissolves easily in water and is a major cause of acid rain, damaging the lakes, rivers, trees and plants. Burning hydrocarbons severely affects the environment; it is a very sensitive issue with the environmentalists, yet no one is taking any real responsibility or addressing any effective solutions.

Oil History

Lacking impervious barriers, oil and gas escapes to the surface, most of which evaporates into the air, leaving behind a black, gooey, tar-like substance composed of hydrocarbons, called bitumen. Evidence shows that throughout history, people in various parts of the world sought out this dense, viscous, sticky tar, and used it in various simple tasks such as waterproofing boats, building roads, building material, burning it for light and even for medicinal purposes.

The first oil wells were drilled in China as early as the 4th century. These wells were drilled using bits attached to bamboo poles and achieved depths of up to 800 feet. The oil was then burned to evaporate the water from brine, producing salt. The early ship industry used the thick black tar-like material to waterproof their boats. Also, the streets of Baghdad were paved with tar as early as the 8th century. In Poland, oil was burned in street lamps to provide light in the town of Krosno as early as the 1500s. This oil had an awful smell and produced smoke full of suds when burned, and because of that was never considered a serious candidate to replace oil derived from animal fat, which was used in lamps for lighting. Native Americans collected this oil as far back as the 1400s. They used oil for medicinal purposes, such as creams and ointments, and also as a mosquito repellent.

With the discovery of distillation in 1853, the modern oil era began. Ignacy Lukasiewicz, a Polish scientist, built the first true refinery, "a distillery," and distilled the crude oil into clear kerosene. It soon gained acceptance over the more expensive whale oil alternative. To expand the kerosene production, Lukasiewicz collected the crude oil from hand-dug wells near the town of Bobrka. Later, hand drilled oil pits, 70 to 150 feet deep, supplied crude oil for his business of "distillation of kerosene." A thriving oil industry in Poland was born as other entrepreneurs joined in to produce oil.

In the 1850s, George Bissell, a New York lawyer, was on a quest for a more reliable source to produce oil commercially for illumination. He believed that he could drill a well to extract the oil from the ground, using the same technique used in drilling for brine and drinking water. Bissell was able to get some financial backing and formed the "Pennsylvania Rock Oil Company." They hired Edwin Drake in 1857, and using steam-powered equipment, began drilling for oil in the summer of 1859.

The American oil industry was born on August 27, 1859, near Titusville, Pennsylvania, when Edwin Drake and Smith drilled a well to a depth of 69½ feet and struck oil. This was not the first oil well drilled in that region, since oil wells were drilled commonly to find fresh water or salt water, and when the drillers struck oil instead, the well was abandoned. Drake's oil well was the first well drilled for the sole purpose of finding oil. This was perhaps the most important and most significant oil well ever drilled.

Almost overnight, Titusville, Pennsylvania, was transformed from a quiet farming region to an oil boomtown. Speculators from all over the country began to lease neighboring land to drill for oil to make their fortunes and the Venango County oil boom quickly spread throughout both Pennsylvania and New York. During the rest of the century the demand for kerosene, which was used for illumination, and the demand for various lubricants (used as grease to lubricate mill machinery and wheel joints) enabled the oil industry to sustain itself.

The modern oil industry began on January 10, 1901 in Beaumont, Texas. Atop the hill called "Spindletop," the famous Lucas Gusher erupted. "Lucas I," the oil gusher greenish-black in color, rose to a height of 150 feet, spilling out 100,000 barrels of oil per day. This was more oil than all of the other oil fields combined. Up until then Pennsylvania was the leading producer of oil in the United States. Its oil dominance ended that day.

Just like Pennsylvania, the state of Texas had been finding, drilling, and producing oil since shortly after Edwin Drake's oil well had first been drilled. Starting in 1866, Lyne T. Barret drilled near the town of Nacogdoches, in an oil field known as Oil Springs. In 1896 the Corsicana Oil Development Company teamed up with oilman John Galey. They drilled an oil well in Corsicana, Texas. As in Pennsylvania, the oil finds were small. Anthony F. Lucas (with the financing of John Galey) began to drill on top of Spindletop beginning on October 27, 1900. On January 10, 1900, at the depth of 1020 feet, they struck oil. Shortly thereafter, Beaumont became an oil boomtown in Texas. It's population increased fivefold to 50,000 people. Within a year as many as one hundred different oil companies joined in, drilling wells on or near Spindletop.

Before Spindletop, the fate of oil was doubtful, because of the limited supply. After Spindletop, at the beginning of the twentieth century, with the introduction of the automobile, the oil boom was launched full force. Petroleum began to power automobiles, airplanes, ships and trains, and to complete the oil dominance, oil began to replace coal. This increased demand lead to oil booms in Texas and California, and this demand has been sustained to this day. Drake's well near Titusville, Pennsylvania began the petroleum era and Lucas's Gusher on Spindletop propelled Drake's era to the modern-age petroleum era of today.

Oil Production

Extracting oil out of the ground is an expensive and complicated process. Only large corporations are able to undertake this type of operation. The first part of an exploration is to identify a favorable site for oil production. Oil exploration's primary objective is to identify oil reservoirs. During the early days, exploration would involve identifying visible features of oil such as oil seeps. Existence of these features served to identify the presence of oil and oil wells were drilled there.

Today's exploration depends on sophisticated technology to establish if hydrocarbons are present. Not only can today's technology predict with reasonable accuracy the presence of hydrocarbons, it can also estimate the amount of hydrocarbons there. During the exploration process, the study of "sedimentology" determines the possibility of organic sediments and the study of "thermal maturity" is addressed to determine the types of kerogen present. Lastly, "reflection seismology" is used on the areas that are identified as good prospects to contain hydrocarbons. Reflection seismology maps the subsurface rock formations and is used today extensively in oil and natural gas exploration, on land and in the depths of the oceans. Geologists analyze the data to determine the presence of traps that could hold petroleum and gas. Once a favorable site is identified, a well has to be drilled to confirm the existence of oil.

The first stage, the extraction process, is to drill a well. If oil existence is confirmed, the oil production will last many years. A successful well is when either oil or gas is found. If oil or gas is not found, or if the quantity of oil is not at a desirable amount for production to begin, then the oil well is referred to as a "dry hole." If successful, however, other oil wells will be drilled to achieve maximum production. The first oil will likely come to the surface under its own pressure, its "natural lift."

Natural lift oil recovery is the most cost-effective way to bring oil to the surface. The underground pressure, which is partly supplied by the presence of gaseous hydrocarbons in the same hole, forces the oil to the surface. An array of valves is connected to the wellhead. The valve array is used to fill the trucks that move crude oil to refineries, where the crude is processed into the products everyone is familiar with. In addition to controlling the oil flow, the valve array prevents oil from escaping into the environment. Natural gushers, while nice to look at, were dirty, and today are a thing of the past. Natural lift production is called a primary production method and accounts for about 20 percent of oil being recovered.

When enough oil is removed from the ground, the pressure subsides. It becomes insufficient to push the oil to the surface. To recreate the natural lift, secondary oil recovery methods are used. These techniques usually involve the drilling of strategically positioned secondary wells and tactically injecting gas, air, carbon dioxide or water back into the reservoir, thus increasing the pressure and re-creating natural lift. Eventually, not even these attempts can re-create a sufficient natural lift and the oil has to be pumped out. These pumps run on natural gas or electricity and do not add much to the price of oil production. Primary and secondary oil production recovers approximately 25 to 35 percent of existing oil.

When secondary oil production is no longer efficient, tertiary oil recovery begins. The most common form of tertiary oil recovery is injecting steam into a well to reduce the viscosity of oil. Oil then flows easier toward the pumping wells, where the pumps bring the oil to the surface. Tertiary recovery methods are expensive and are used only when the price of oil justifies it. When the price of oil is low, the wells are capped off. When the price of oil justifies it, the oil wells are brought back into production. Tertiary recovery is capable of recovering another 5 to 15 percent.

Refinery

Once crude oil is recovered from the ground it is transported to a refinery. Crude oil consists of a mixture of hydrocarbons. Simple distillation is used to separate the crude oil into its different components. Different products have different boiling points. Naphtha, having the lowest boiling point, is recovered first; jet fuel, kerosene and diesel fuels, the middle distillates, are recovered next, and finally the residual fuel products are recovered at temperatures in excess of 1000 degrees Fahrenheit (°F).

A simple distillation procedure is the first step in the refining process. Simple distillation provides around 20 percent of the lighter desirable product naphtha, to produce gasoline, and as much as 50 percent of the residuum, requiring a more extensive process to produce desirable products. For example, the use of a catalytic cracker converts oils with high boiling points to fuels having lower boiling points, producing more desirable products, mainly gasoline. A cooker, on the other hand converts the residuum, the heaviest of oil, by cooking, or using a thermal-cracking process to the more desirable light products. At the end of the complex refining process, the crude oil yields 95 percent of usable products and only 5 percent residuum. Out of every barrel of oil, 29.6 gallons, or two-thirds, is in some form of gasoline and diesel fuel. In percentage terms, the quantities are as follows, gasoline (47%), distillate fuel (24%) and jet fuel (10%). This ratio provides for 81 percent of oil in every barrel to be burned somewhere in the transportation industry as fuel and leaves 19 percent for all other products, such as liquefied and still gases, kerosene, residual fuels, lubricants, waxes, petroleum coke, road asphalt and other miscellaneous goods.

Although oil is produced in various parts of the world, the refining itself is done locally in the consuming areas because it is easier and cheaper to transport crude oil than it is to transport the finished products. Additionally, local refining makes it easier to respond to the seasonal shifts in demand. The U.S. refining capacity is about 16 million barrels per day, demonstrating that the capacity of

U.S. oil refineries has not kept up with the ever-increasing oil demand. Low profit margins and low oil price is blamed for that. As of 2006, capacity utilization of U.S. refineries is at the maximum, and with the U.S. consumption of 21 million barrels of oil per day, Americans have to look elsewhere not only to import crude oil, but also to have a substantial part of the oil refined into individual components, thus adding considerably to the overall cost of oil products.

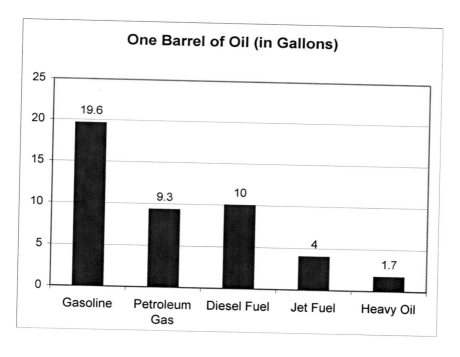

How much Oil is in United States?

As of 2006, United States oil reserves are estimated to be about 30 billion barrels. Compared with the Middle East estimates of some 750 billion barrels, these reserves are miniscule and insignificant. A more telling statistic is that the U.S. consumes some 7.5 billion barrels of oil each year, thus the existing thirty billion barrels would provide less than a five-year oil supply for America. It has been determined that the wildcats have pretty much drilled throughout the lower forty-eight states and there is no more inexpensive oil to be found. There may be a few gallons discovered here and there but that's about it. The land production of the United States is currently just over 1 million barrels per day.

At present, the U.S. oil companies are looking at offshore drilling opportunities for new oil discoveries. As the land oil production dwindles to just about nothing, the gulf coast, the east coast and the west coast of the United States are being explored more extensively than ever. The biggest offshore oil production is currently in the gulf coast region. In order to reach these oilfields, the drilling is being carried out in water depths in excess of two miles.

A disturbing statistic, however, is the fact that the oil production in the Unites States peaked in the 1970s and has been declining ever since. Everyone in this country has largely ignored this fact. The U.S. oil producers cannot keep oil supplies constant due to the fact that it is increasingly harder to replace the old depleted oil supplies with the newly discovered ones.

Even adding Alaska's ANWR to the overall outlook would not reverse the U.S. oil decline slump. Alaska's ANWR is the subject of constant conflict between environmentalists and lawmakers. Those who want to tap the ANWR reserves think that those oil reserves would in some way reverse the decline and solve the U.S. oil dependency on foreign oil. The fact is that by the best estimates and assumptions, the ANWR contains about 10 billion barrels of oil, which is a relatively insignificant supply. With the U.S. consuming 7.5 billion barrels yearly, these reserves could satisfy the country's oil needs for about a year and half. Additionally, tapping the ANWR oil reserves would require initiative, which could last more than ten years before the first oil would flow. Alaska, for now, remains a mystery. The only sure way to find out if and how much oil is in the ground is to drill a well, and as long as the environmentalists get their way this will not likely happen.

Understanding the places that have been explored and the few remaining potential places that remain to be explored, and taking into consideration the capabilities of today's drilling technologies; assumptions can be made that the new easy way to produce crude oil (the liquid oil Americans have been accustomed to) will not come from the USA or their coasts. So what else is out there that will satisfy America's appetite for oil?

United States and Unconventional Reserves

The United States and its northern partner Canada have large amounts of oil reserves left. These reserves are the unconventional reserves. They are: oil still left in the ground, the Tar Sands of Canada and U.S. oil shale. Even though today's oil prices keep these reserves out of reach, they will play a role in the future. These reserves are huge and quite comparable to those of the Middle East. Therefore they need to be mentioned.

Oil left in the Ground

Just because the well has run dry does not mean there is no more oil in the well. In fact, there is still plenty of oil remaining. By some estimates some 50 to 70 percent, possibly more, is left, which provides an estimated quantity of 200 to 300 billion barrels of oil remaining. To put it in perspective, these reserves are ten times as large as the proven oil reserves that exist in the United States.

It certainly sounds optimistic, to hear that more than half of the oil is still out there. The problem is there is no easy economical way to bring the oil to the surface. In addition, the oil drillers lack the technology that is able to locate the oil within the oil reservoirs. Eventually though, the oil-producing companies will discover new ways to get to these reserves. However, it may be at much higher costs.

A second problem is how much energy has to be used up to produce one gallon of oil. If oil producers have to use up one gallon of oil to recover one gallon of oil, they are not staying ahead of the game. Oil energy is used to power the pumps to recover the oil. It is used to power the trucks which take the oil to the refineries. It is used by refineries to produce gasoline. And trucks use this fuel to take the refined gasoline to gas stations to be sold. Therefore, until new cost and energy-efficient ways are found, this oil will remain hidden in the ground.

Oil Sands of Athabasca, Canada

The Oil Sands (Tar Sands) are exactly what the name implies. This heavy tar-like oil is mixed with sand. These heavy hydrocarbons tend to be very thick at room temperature and are usually much harder to recover out of the ground than the lighter crude oils. Oil sands are currently found in many countries; however, the largest oil sands deposits in the world are found in Alberta, Canada and Venezuela. Currently, the Athabasca Oil Sands of Athabasca, Canada, has an estimated 300 billion barrels of technologically retrievable oil. It projects an additional 2 trillion barrels lying in place, waiting for new technological advances to be recovered. Canada produces a fraction of 1 million barrels of oil per day from these oil deposits, and while most of this production finds its way to the United States, the production would need to double several times before it could become a major part of U.S. consumption.

The bitumen deposits, which lie near the surface are being recovered by open-pit mining. Huge loaders and trucks, some with capacities of up to 380 tons, are used for these operations. These trucks transport the oil sand to the processing plant, where hot water is used to separate the bitumen from the sand. This heavy oil has to be diluted with other lighter hydrocarbon products to make crude oil, which is then shipped to the refinery. One ton of tar sand yields about 25 gallons of bitumen.

When the bitumen deposits are buried deep below the ground, surface mining is replaced by in situ techniques. Steam Assisted Gravity Drainage is the widely used technique. First a series of horizontal collecting wells are drilled near the bottom of the deposits. Then the steam is injected a few feet above these wells. The steam heats the thick oil, reducing its viscosity. This oil then flows to the collecting wells below, where it is recovered and pumped to the surface.

Tar sands production is sending out huge amounts of greenhouse gases. It is a major contributor of the earth's atmosphere pollution and is responsible for global warming. Oil sands are currently the dirtiest way of obtaining energy, but as long as the United States continues to need this oil, the environment may be ignored.

Canada has found that as production increases, the cost of production decreases dramatically and the oil recovery from these fields is becoming attractively cost-effective. As the price of crude oil increases, the tar sands production is expected to grow dramatically. Hopefully, a small part of this newly acquired revenue will find its way to helping prevent the dirty greenhouse gases that are entering the atmosphere.

United States Oil Shale

In addition to the 300 billion barrels of oil still in the ground, the U.S. has large reserves of oil shale deposits. These are located in the western part of the United States, Wyoming, Utah and Colorado. It is estimated that these deposits contain more than 2,000 billion barrels of oil. This is more oil than all of the world's liquid oil reserves combined and dwarfs the 700 billion barrels of known oil reserves in the Middle East. One ton of Green River oil shale, for example, yields approximately thirty gallons of kerogen oil.

The problem of how to produce oil efficiently and economically prevents production. The oil shale does not actually contain oil. The shale rock contains organic matter, called kerogen, which must go through a heat process called distillation to produce oil. At temperatures of 450°C to 500°C, the kerogen is separated from the shale, producing oil.

Oil shale is not competitive with crude oil below $40 per barrel. Production of oil from oil shale gained interest in the 1970s, when the price of oil temporarily reached $40 per barrel. Later, as the price retreated below $40, so did the interest in oil shale. Oil prices must stay high permanently for the interest to return. Unstable oil prices will prevent returning interest. Consider this: a higher price of oil attracts new exploration and additional production elsewhere. This phenomenon in turn provides a new glut of oil worldwide and causes the price of oil to drop, thus making it uneconomical to produce oil from oil shale. In the 1970s, investors learned of this phenomenon the hard way when after large investments in oil shale technology, all of them, one by one, went out of business. And not surprisingly, due to the same phenomenon, none of these investors returned.

Just as the oil sands of Canada, the production of oil from oil shale creates large quantities of green house gases. Are the oil companies going to get a pass from the environmentalists? Perhaps in the future, when liquid oil reserves have been all but depleted and much cleaner ways of production have been found. Only then may the oil from oil shale find its way to the gas tanks of American drivers. However, don't bet on seeing these oil reserves any time soon.

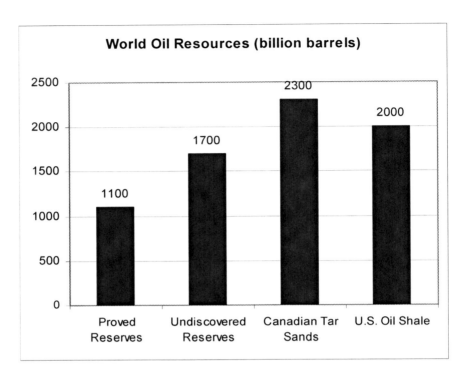

This illustration shows that by combining the 2,300 billion barrel reserves of Canadian Tar Sands and the 2,000 billion barrel reserves of U.S. Oil Shale; the total dwarfs the 750 billion barrel reserves of the Middle East. One day the new Middle East could well be the North American continent.

Oil Supply vs. Oil Demand

Oil companies are aware of places they are tapped in. The Middle East, which is the largest reserve found so far, is being siphoned out as if there was no tomorrow. The countries with the largest reserves are Saudi Arabia, Iran,

Iraq, Kuwait and United Arab Emirates. A few smaller oil reserves exist in Russia, Nigeria, Libya and Venezuela; but baring any new major discoveries in other parts of the world, these countries will dominate the world's future oil production. These countries will supply the world with most of the needed oil in the future. Numerous countries with much smaller oil reserves account for a small portion of the total, but none of these provide any meaningful amounts for the ever-increasing demand.

By most estimates, the world now has 50 to 60 years of oil left. The total world reserves are estimated at 1,100 billion barrels. The Middle East accounts for the majority of the world's supply with around 750 billion barrels. Venezuela and Russia are estimated to have 70 billion barrels of oil each and Libya and Nigeria are estimated to have 35 billion barrels of oil each. In all, these nine countries hold 950 billion of the 1,100 billion barrels of estimated reserves. This leaves 150 billion barrels for the rest of the world.

What about the oil demand? According to BP, in 2004, the United States oil consumption was 20.5 million barrels per day and is expected to keep increasing. According to EIA, by 2025 the consumption in the U.S. is estimated to reach 27.3 million barrels per day. The world's demand will increase from 80.7 million barrels per day in 2004 to over 119 million barrels per day by 2025. The largest increase in consumption is projected to be in emerging economies such as Asia, which includes China and India.

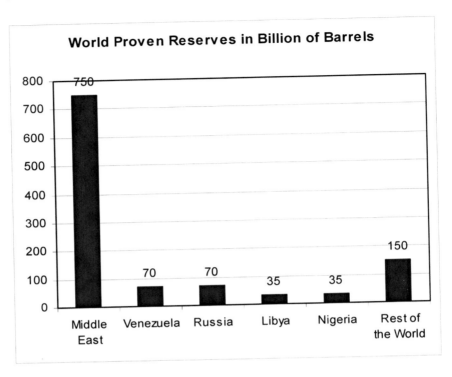

As of 2006, most of the world producers are pumping their maximum capacities. So is it realistic to expect the consumption to increase by 50 percent? Especially, when one takes into consideration that new discoveries have not kept up with increased demand for many decades. One day in the near future, oil production will tip the scales and start declining; when the world learns that oil production is on the decline and the consumption continues to increase at an unprecedented rate, the other shoe will drop and the price of oil will show it.

Other Oil Theories

There are the pessimists out there who believe that most of the available oil has already been found. They argue that the oil supply is finite and they rely on information from past discoveries to conclude that all the large reserves have already been discovered and that future discoveries will be small. They suggest that some countries of the world overestimate the actual oil reserves and that these reserves are in reality much smaller than the current estimates show. Thus, the statistical conclusion of oil pessimists is that the inevitable oil bust is near. The most profound pessimists out there believe that the world's oil production has already peaked or is going to in the next few years. These alarmists predict imminent oil crisis and paint a picture of worldwide depression.

The pessimist's use Hubbert's peak theory as a model to make the case that the crisis is inevitable. In 1956, Marion King Hubbert, a geophysicist and geologist, created a model of known reserves and predicted that U.S. oil production would peak in 1970 and that world oil production would peak in 2000. The U.S. oil production peaked in 1971; accurately predicted by Hubbert's theory.

Hubbert created a mathematical model which predicted that the maximum oil production of an oil field would follow a bell-shaped curve (Hubbert's curve). The phenomenon can be explained as follows: When oil is first discovered, production is slow. As a new infrastructure is built, the production will increase. Eventually the oil field will reach the maximum oil output and production will start to decrease. At some point before oil is completely depleted, it will become uneconomical and that oil field will be abandoned.

There are a number of other organizations that provide oil peak predictions. Colin Campbell (geologist) of the Association for the Study of Peak Oil and Gas had predicted an oil peak would occur in 2004. He defends Hubbert by saying that events which happened after Hubbert's prediction delayed the peak. The energy crises of 1973 and 1979 and the 1980 and 1990 recessions have slowed down demand; thus pushing the estimated peak into the future. Campbell argues that there are no large new oil fields to be found to prevent the upcoming energy crises and that the upcoming peak will result in a catastrophic worldwide economic depression.

The organization ASPO predicts that oil production will peak around 2010. The United States Geological Survey predicts that there is enough petroleum to maintain current production levels for the next fifty to one hundred years, in which they predict an oil peak in 2037. The Energy Information Administration, along with the International Energy Administration, predicts no peak before at least 2025. To sum up, in a pessimist scenario, the implication of a world peak is: shortages of energy, large runaway inflation, exploding transportation costs and lower living standards worldwide.

The optimists, on the other hand, share an idea that any crisis is highly unlikely. They believe that many of the oil reserves still have not been discovered and that reserves, larger than the ones in the Middle East, are still to be found. In addition to the undiscovered oil, they believe that as the price of oil increases, and with improvements of existing technology, the past oil fields will be revisited and what is still left in them will be extracted. Finally, they believe that the vast oil shale reserves will sustain the current demand, shortly after the price of oil justifies it. Optimists estimate that less than half of the oil has been found to date and that some 2 trillion barrels of oil are yet to be discovered. They believe that offshore drilling will produce massive discoveries as technology allows drilling in much deeper waters. Optimists' point of view is that the oil production peak is not even close.

An interesting idea of some optimists is that the oil is a "renewable resource" (Abiogenic theory of petroleum origin). The idea proposes that petroleum is formed deep within the earth's crust. It proposes that large carbon deposits existed from the time when the earth was first formed. These hydrocarbons migrate toward the earth's surface through cracks in the crust, primarily in the form of methane. The idea claims that the oil reserves are never depleted because they are constantly replenished from depths below. The current theory allows for just a limited window of depth for hydrocarbons to form; concluding that a certain range of temperature and pressure is required. Abiogenic theory suggests that large quantities of hydrocarbon deposits are to be found much deeper. They suggest that based on the rate of seepage of today's hydrocarbons, the earth has sufficient enough quantities of hydrocarbons to satisfy the world's energy needs for a million years. They maintain that the theory of "oil peak" is a fraud, and a conspiracy of large oil companies to raise prices.

Most optimists, though, assume present day acceptance of hydrocarbon formation, which claims that organic matter, the remnants of dead plants and animals, is responsible for today's hydrocarbons. However, their most extreme case claims are that the oil supplies will not run out for at least another couple hundred years and by then the technology will be advanced sufficiently enough to provide the world with an alternative energy source. Argument continues to leave everything to the law of supply and demand, which claims that while today's price does not justify new exploration, oil price increases will enable new oil exploration in remote locations of the planet, thus new, plentiful discoveries.

Secondly, the argument continues that the huge unconventional oil reserves are going to extend any potential oil peak far into the future. Again, emphasizing that the price of oil will determine when these oil reserves are going to be tapped. Hence, ironically, the theory advocates that higher oil prices will solve the oil supply shortages. They agree that a peak is going to be reached eventually and that the oil supply will start a gradual slow decline, but rather than bringing forth the doomsday scenario of the pessimists, they insist that the gradual decline in oil supply will contribute to completing the transformation of the U.S. economy, from petroleum to alternative fuel, without any crisis developing.

No one will argue that optimist's point of view is not exciting. The idea of not having to do anything today is certainly lucrative. If there indeed is a two hundred year window to find an alternative fuel, let's not panic and get it right.

And so who is correct? Pessimists see the upcoming oil peak and the resulting higher prices as an open invitation to inevitable economic crises, shortages and runaway inflation. Optimists see the rising prices leading to the oil peak as an opportunity to increase oil exploration and an opportunity to transfer from oil based economy to alternative fuel based economy. Neither side argues that an oil peak, a point at which the existing demand exceeds the existing supply thus creating a scenario of rising prices is not going to occur; rather, what is being argued is when this scenario is going to happen and what should be done about it.

To conclude, oil is going to stay here for quite some time and it is up to the U.S. government to decide what path to take. It is true that markets will eventually resolve the oil supply-demand on their own, but during that time a wave of large price swings and shortages may occur, reinforcing the pessimist's scenario of a catastrophic worldwide economic depression. No one can predict with certainty to what extent this type of do-nothing scenario would affect the U.S. economy, but it is hard to imagine that the U.S. government would just stand by and allow this unpredictable pessimist scenario to play out without any intervention.

For more detailed information on any subject in this chapter, go to:

http://en.wikipedia.org/wiki/Timeline_of_evolution
http://www.eia.doe.gov/kids/energyfacts/sources/non-renewable/oil.html
http://www.sjgs.com/history.html
http://en.wikipedia.org/wiki/Crude_oil
http://eia.doe.gov/pub/oil_gas/petroleum/analysis_publications/oil_market_basics/

http://en.wikipedia.org/wiki/Oil_shale
http://en.wikipedia.org/wiki/Abiotic_oil
http://www.du.edu/~jcalvert/phys/carbon.htm
http://www.hubertpeak.com/hubbert/1956/1956.pdf
http://en.wikipedia.org/wiki/Hubbert_peak_theory
http://en.wikipedia.org/wiki///CO2

2.
Oil War Games

Cost of Oil Imports

During the 1973 energy crisis, the price of oil jumped from $2 to $10, and in 1979, during the Iran-Iraq war the price of oil increased from $10 to $30 per barrel. The cost of imported oil is forever increasing, in which millions have turned into billions. Currently, the cumulative cost of imported oil is being described in trillions of dollars. To illustrate, with imports of 14 million barrels per day, 365 days a year and assuming an average price of oil of $70 per barrel, one year's bill for Americans could exceed $350 billion.

The price of oil is increasing continuously, as are the oil imports, which are projected to surpass 20 million barrels per day by 2025. In the not-too-distant future

Americans will be spending $1 trillion on imported oil every year. It is estimated that it may cost 2 trillion dollars to transform this economy to an alternative fuel source.

A second part of the oil cost that most Americans are not aware of is the cost of the U.S. Army, which America must maintain in order to prevent any potential oil disruption. The army costs are not reflected in the price of oil. Maintaining a larger army means all Americans are in reality subsidizing the price of cheap oil by paying higher taxes. What price will Americans have to pay before the U.S. government realizes that it is much cheaper to switch to a different source of fuel?

It has been suggested that $1 added to the cost of each gallon of gasoline accurately represents a material price tag of war guarding the oil supply; however, there is yet another cost. The cost of the dead soldiers has somehow been omitted from the total. What price tag should be added to the price of gasoline to compensate for these lives? One dollar per gallon, two or ten; the answer depends on who you talk to. Better yet, ask someone who lost his or her son or daughter. Families that have gone through the trauma of losing their child in a battle for oil will say that no price is worth American lives.

Cost of Terrorism

On September 11, 2001, in a secretly organized terrorist attack on the United States, nineteen terrorists changed America and the world forever. Security checkpoints in airports and government buildings are now a way of life. Americans are now treated as if they are part of Al-Qaeda until they pass through the security checkpoints. These attacks killed 2,986 people. How would the United States and the world change if another similar attack occurred?

Many Middle Eastern countries breed terrorists. These terrorists join to form terrorist organizations with a specific purpose. The first purpose is to inflict brutal harm to Americans, and second, when the terrorist groups combine forces with many of the Middle Eastern governments, is to eradicate Israel. The goal is to erase Israel from the global map. A nuclear weapon at the hands of many of these governments or many of these terrorist organizations would certainly accomplish both goals at the same time. At the hands of many of the Middle Eastern leaders, a nuclear weapon would signal the end of Israel where Israel would become a barren desert property.

Similarly, setting of a nuclear weapon in the middle of New York would dwarf the 9/11 disasters. The casualty list would be one hundred, perhaps one thousand times greater. Nuclear weapons at the hands of terrorists would signal the potential of real destruction in which 3,000 fatalities could easily become 300,000 or 3,000,000. Having a nuclear capability, these nations would become the world's "superpowers," making dealing with them agonizing, even for the United States.

Inflicting pain to the U.S. is of real concern. The U.S. now has close to 150,000 soldiers in Iraq. Americans should be able to safely conclude that this force is capable of resolving any situation. Nevertheless, the U.S. Army has a difficult time controlling a few thousand terrorists. The U.S. Army is finding out that the overwhelming U.S. weapon's superiority is no longer effective in fighting this type of war; thus acknowledging that the war against terrorism is much more difficult then previously thought.

What if the terrorists were successful and got hold of a couple of nuclear weapons, and instead of "traveling" to the United States, they decided to the destroy oil fields in Saudi Arabia? The Saudis would certainly have an "out of business" sign on their front door and the United States would face a shortage of 1.5 million barrels per day. One nuclear bomb would turn off that country's oil spigot. Of course, the rest of the world would not fair much better. Saudi Arabia pumps 10 million barrels per day, and supplies oil to many other countries. Therefore the fate of quite a few countries throughout the entire world would mirror the U.S. tragedy.

Currently, the U.S. is facing the most difficult terrorist problem yet. Is the United States going to be able to leave Iraq in the near future or are they committed to stay for a long time? If the United States withdrew from Iraq, thousands of terrorists would be out of work. These terrorists would not be satisfied collecting unemployment. Now they are busy assassinating Iraqi and American soldiers, but in the absence of American troops it would not take these fanatics long to find new jobs, such as blowing up oil fields in their own back yards and all over the Middle East. Everyone is familiar with the terrorist's thinking, which is to kill and hurt Americans and American interests everywhere they can. Disrupting the oil supply, which Americans rely on, would certainly be one of the goals. Threatened oil supply is probably the most likely reason that the United States will have to stay in the Middle East for a long time to come; to keep the terrorists busy so they do not have time to cause any more destruction elsewhere.

Because the U.S. needs the Middle East oil, the U.S. Army is being used to keep the region stable. What would the wars in the Middle East look like if there was no oil? First, the crazy fanatics such as Iraq and Iran would have no money to buy sophisticated weapons for their use, nor would they have money to supply arms to the terrorist factions such as Hezbollah and Al Qaeda, who are determined to destroy Israel.

Secondly, the U.S. would not need to dance around the issue of what to do if one of these countries began to develop nuclear weapons. Today it is "diplomacy in disguise." The U.S. negotiates so they do not offend other Middle Eastern countries. The United States and the rest of the world fear retaliation in the form of oil embargoes. Without the fear of this kind of retaliation, the appetite for nuclear weapons by these countries could be suppressed just by air bombing the newly discovered nuclear facilities without any U.S. casualties.

Because the U.S. is sort of the policeman in suppressing these conflicts, they are always leading these fights. The U.S. Army is almost always at the center of all the Middle Eastern battles and therefore the U.S. is always the most hated nation throughout that region, and that is unlikely to change anytime soon. Perhaps it would suffice if the United States was not dependent on the Middle Eastern oil. It would be reasonable to assume that without the worry of oil embargoes, the U.S. could control the nuclear friction in the Middle East by air, without any ground troops whatsoever. By not needing any Middle Eastern oil, the United States would continue to be hated by the Middle Eastern counties, but without a nuclear threat or any dead soldiers.

Since 9/11, Americans have been sort of lucky. The Afghanistan war and the Iraq war have confined the terrorist concentration within the Middle East and the war has been fought away from American soil. Still, that is unlikely to last forever. For now, terrorists fight the American Army stationed in the Middle East, and as long as they believe they are succeeding, inflicting casualties, damaging the U.S. Army and breaking down the army's moral, they will continue to fight there. Once these militants realize that this strategy is no longer effective, then to gain better success, the strategy will shift toward different tactics. This means resumption to the old strategies of attacking American civilians and American interests in other countries. Inevitably, this strategy will bring these radicals back to American soil.

As long as the U.S. continues to be dependent on an oil supply from other parts of the world, terrorism will continue to be a very effective weapon against the United States interests and the U.S. Army will have to guard against terrorism not only within the U.S. borders but also throughout the rest of the planet. Fighting terrorism on many fronts in many parts of the world will extend the U.S. Army to be thinly dispersed everywhere. This includes the United States borders. Thinly guarded American borders and oil interests will provide easy targets for the terrorists. This extreme exposure to terrorism means that many more Americans will be massacred in the name of oil.

Oil War Games

Cost of Blackmail

Blackmail must be on the minds of the U.S. government each and every day. The U.S. President and the U.S. government know well who owns the oil. They know it is the countries in the Middle East, who are no friends to the United States. For now, while oil surpluses exist, the Middle East allows the U.S. to buy their oil; however, once the excess capacity subsides, the Middle Eastern oil, including the Saudi's oil, is likely to find its way to other parts of the globe. Doing business with China, India and other countries in Asia with their ever-increasing demand, will be preferred over doing business with the United States by most of the Arab nations.

Venezuela has also been unfriendly to the United States since the formation of OPEC. For now the U.S. is a valued customer, but that would change very quickly if Venezuela found another buyer. When the Asian demand creates a supply squeeze, then just as the Arab oil supply, the Venezuelan oil will find its way to the far east. Up to now, Venezuela has been talking about disrupting the supply of oil to the U.S. very quietly. However, Hugo Chavez is currently talking about oil disruptions to the U.S. more openly and more frequently than any other time in the past. One day in the future this dictator will snap, and because the U.S. imports 1.5 million barrels of oil per day from Venezuela, the U.S. will have to act.

Lastly, what if OPEC suppliers decided to sell their oil to China instead of the U.S.? Or what if the Arab countries decided to keep the oil for their own consumption? Or what if they were forced by terrorists to interrupt the oil supply to United States? Would the U.S. start a World War

III? The United States is fully aware of the oil production capacity in the world and knows that as of 2006, there is no extra production capacity left to replace any possible oil supply disruption. Therefore any oil disruption or another oil embargo is just unacceptable.

In conclusion, oil embargoes and blackmail demands will become more common as oil supplies shrink. What's more, as long as the United States continues to stand by Israel against the Arab world, terrorism and blackmail will be a part of American life. Oil shortages and sharp price spikes will be a cost Americans will have to endure. Thus again, the U.S. being a self-reliant nation is the only solution to blackmail.

The Cost of War and its Consequences

War is a messy business in which people die. History will no doubt provide thousands of pages speculating the reasons why the two Gulf Wars were fought. Historians and scholars will provide their conclusions based on their own personal reasoning of facts. The overwhelming fact will remain that the Gulf Wars were oil driven. The U.S. government may provide multiple reasons for the war, be it weapons of mass destruction or providing freedom to oppressed countries, but the facts will not change; the facts will remain facts.

The events that took place in these conflicts could be oversimplified and described as follows: The first Gulf War in 1991 was fought in response to Iraq invading Kuwait on August 2, 1990. Iraq, who just finished an unsuccessful bid to control Iran and their oil fields, decided to take control of Kuwait, a much easier target, to increase the length of its border with the Persian Gulf and increase its oil production capacity. After Iraq's successful invasion of Kuwait, the U.S. saw Iraq with too much power over the control of oil and decided to restore the Middle Eastern oil stability by restoring free Kuwait. To accomplish this, the United States established a coalition of many countries to drive Iraqi forces from Kuwait. Once the Iraqi army was defeated and forced out of Kuwait, the war was over.

The second Gulf War (2003) was fought over much the same premise. It was to restore the gulf's stability by preventing Saddam Hussein from acquiring overwhelming weapons of mass destruction, with which the oil region could again be threatened. The threat to the United States was again oil disruption. The toppling of Saddam Hussein and the gain of Iraqi freedom was a "welcomed appreciated bonus" of that oil war.

Officially the cost of these conflicts were 345 dead and 100 wounded in the 1991 war and over 2,850 military dead and over 19,350 wounded (and counting) as of August, 2006. The government and the media always downplay the wounded. The wounded soldiers somehow provide for an easier war outcome; nevertheless, once these soldiers lose a part of their body and become permanently crippled, their outlook on life changes permanently. Americans must be made aware of the fact that being wounded does not just mean being treated for some type of scratch, or a cut needing a band-aid and in two days the boo-boo is gone.

The word "wounded" may mean having had the guts splattered all over the desert sand. It may mean permanently mentally disturbed, rather ill, and not able to hold on to any job – not even the simplest job – to support oneself or one's family. It may also mean permanently disabled, disfigured or mutilated, with gruesome, hideous, and horrific wounds to the body; it may mean the loss of an arm or a leg or even becoming paraplegic and permanently confined to a wheel chair, unable to take care of even the most basic functions. And it may mean horrid, repugnant, and ghastly burns all over the body or a horribly shocking disfiguration of a face burned beyond recognition. Whatever the injuries, these soldiers will have to live with them for the rest of their lives.

When Americans understand what the word "wounded" really means, the war casualty list will take on a whole new meaning. The 2,850 dead and 19,350 wounded will become 22,200 casualties of war, and these statistics paint a much different picture, especially if the 9/11 casualties (the dead and wounded) were added to the total. One day Americans will realize that with this many American victims and this many casualties of war, the cheap oil they get in return is not worth the price. An America self-sufficient in energy is an attractive alternative. In fact, achieving self-sufficiency is the only course of action that will prevent American soldiers from being mutilated, disfigured and slaughtered.

The Cost of Government Friction

Guarding the oil reserves abroad creates tensions; not only between nations, but also within the U.S. borders. The U.S. government remains divided across party lines. Should the United States be involved in these oil conflicts? While one party sees resolving these conflicts, which threaten the oil supply as a national security issue, the other party disagrees and contemplates exactly the opposite. While one side argues the consequences of doing nothing far more dangerous than involvement itself, the other side again disagrees and argues a better way may exist somewhere. The fact is that both sides need to be held accountable by Americans; it is not just one party's war. Getting involved threatens soldier's lives, doing nothing threatens a disastrous economic collapse, and so unless both sides handle these difficult issues on bipartisan bases, no one wins.

When the subject at hand involves war or national security, the U.S. government should not be divided along party lines, as is currently the case. One party stands for the war and the other party stands against the war. It should not be the hope of something going wrong with the war that can then be used in the next election. It sure seems in this case that the terrorists won big time. They have managed to transform a resolute unified country after 9/11 to a country divided along political lines. It has reached a point where some in the U.S. government are even suggesting accepting defeat and to cut and run.

If anyone is wondering what may occur if the U.S. retreated to terrorists in Iraq, just look at what has happened between Israel and Hezbollah. Although Israel totally leveled the Hezbollah strongholds in southern Lebanon in 2006, Hezbollah is viewed as a victor in that conflict. Hezbollah was not totally eradicated, and because they were able to withstand the bombing of an Israeli army, they are now treated as heroes in the Arab world, despite the fact that they have directed the fight strictly against civilians in northern Israel. Despite the fact that Hezbollah has been hiding among civilians in their own country, thus putting their own people in danger of being killed, of which many were. A cease-fire will now enable Hezbollah to rearm, thus able to fight another day.

Drawing a parallel between these two conflicts, one may conclude that withdrawal from Iraq prematurely would give the terrorists huge moral victory, no doubt elevating them to royalty status in the Arab world. The increase in recruitment to become a terrorist to hurt Americans could be unprecedented if these radical terrorists witnessed that even the mighty American Army is incapable of dealing with them. On the other hand, one may no doubt conclude that retreat of U.S. Army from Iraq may end hostilities. You decide what is likely to be the case.

The inability and incompetence of the two political parties to provide a unified solution to the oil conflicts cause unwarranted tension within the U.S. population. The population imitates their individual parties and rather than see the war for what it really is, the death, misery and destruction, Americans see the war as party politics. What an effective public relations concept conducted by these politicians. Americans are so busy following and supporting their sides that they do not even realize this country is at a deadly war where Americans are being killed. Thus all Americans listen to the hate speeches of the government officials, the movie stars and many activist organizations and get sucked up into the game of dirty politics. The U.S. government has many important issues to deal with and could do a much better job with those issues if they were not preoccupied by the oil wars.

Economic Instability

Recall the fivefold increase in 1973 and the long gasoline lines and panic throughout this country. Those were created by shortages of just 5 percent. In 2006, the U.S. imported approximately 1.5 million barrels from Saudi Arabia. The loss of the Saudi's oil supply would create a 7 percent shortage. What would happen to the U.S. economy? What would happen if the gasoline at the pumps increased fivefold overnight? Incidentally, that would increase the price of gasoline from $3 to $15 per gallon. Would the United States survive without a problem; would they merely experience some form of mild recession or would the U.S. economy be on the brink of economic collapse? Americans do not like uncertainty. They depend on their government to resolve these problems, and when the problems occur, Americans point fingers at their government and a Monday morning quarterback of what could have been done.

It is certain that sudden substantial oil price hikes would not only have consequences at the U.S. pumps but also at the U.S. stores. America depends on the trucking industry to distribute the bulk of the domestic goods across the country and the trucking industry depends on inexpensive fuel to provide these services economically. Substantially increased fuel costs to the trucking industry would certainly translate to increased prices at the end of a line, the U.S. stores. Fresh fruits and vegetables, for example, which are transported from coast to coast, would certainly be sold at a premium.

A fivefold increase in the oil price would certainly cripple the U.S. automotive industry. The sales of large automobiles would screech to a halt and large inventories of unsold vehicles would slow down, perhaps even stop the production of new autos altogether. And with one of six jobs in the U.S. related to the automobile industry, the fivefold jump may cause a lot of unexpected layoffs.

The sudden spike of oil would slow down travel as well; not just because the travel costs became more expensive, but as layoffs became more frequent, less people would be taking trips. The lack of travel would slow down the airline industry and vacation destinations and hotels would experience a much slower business, compounding to the layoffs.

This domino effect of the above scenario exists because of the continued oil dependence on foreign nations. It is not an over-exaggerated scenario and because of the world's instability this can happen at any time, virtually overnight. The U.S oil reserves could technically provide this country with oil for several months but they would not prevent prices from skyrocketing, as many claim. Use of these reserves would only delay the inevitable. Eventually, these reserves would need to be replaced, which would ironically make the situation worse. Replacing these reserves would generate additional demand, creating additional upward pressure on oil prices in a world of limited supply.

A question should be asked, "If the U.S. is unprepared, and a sudden price spike of five or tenfold occurs, (a price jump to $15 to $30 per gallon), how long could this price hike be absorbed by the American businesses and the American consumer before causing permanent damage to the U.S. economy?" Thus the same conclusion, eliminating oil dependence, would provide for a more dependable economy not subject to the world oil instability.

Oil War Games

Future Oil Prices

For those who think a fivefold increase in the price of oil, from $3 per gallon to $15 per gallon, is outright nonsense, then consider these examples taken from U.S. history.

In 1859, when oil was first discovered, the price of a barrel of oil was as high as $50, yet a few short years later the glut of the oil supply has brought the price of oil as low as 10 cents per barrel. A sharp increase in oil supply had resulted in a 500 percent increase in price. For some one hundred years, the price of a barrel of oil has remained in single digits. As recently as 1972, it was around $2. Then in 1974, it jumped to $10 per barrel and went all the way to $35 per barrel in 1980. The price had hovered between mid-teens and $30 per barrel until 2004, and in 2005 the oil has seen its price hit $70 per barrel. Where exactly is the price going?

In the past thirty-three years the price of one barrel of crude went from $2 to $70, which means it has gone up thirty-fivefold, or better yet, it has doubled five times. It has doubled once every seven years for the past thirty-three years. It is like the penny story. Take a penny, double it every day, and in less than a month you will be a millionaire.

If one believes that "past behavior is a good predictor of future behavior," as Dr. Phil McGraw, (of the Dr. Phil Show), says frequently, then everyone can agree that the future price of oil will continue to rise in a similar manner. This means that the price of a barrel of crude oil will double on the average of about every seven years.

Starting in 1972 with a $2 price of crude oil, extending the past thirty-three year behavior thirty-three years into merica: The Oil Hostage

by the year 2040, the price of a barrel of
,048. The price of gasoline will reflect a
which it will reach $64 per gallon.

f oil & the price of gasoline doubling every 7 years.

YEAR	COST OF BARREL OF OIL IN DOLLARS	COST OF GALLON OF GAS IN DOLLARS
1970	2	.33
1977	4	.50
1984	8	.75
1991	16	1.0
1998	32	1.5
2005	64	2
2012	128	4
2019	256	8
2026	512	16
2033	1024	32
2040	2048	64

The price went up thirty-fivefold in the past thirty-three years. Who can prevent it from going up thirty-fivefold in the next thirty-three years? At what price will the U.S. economy crash because it cannot sustain itself? Will the price behave as a perfect model stock would, having higher peaks and higher valleys? In the summer of 2005, the price of gasoline went to $3.50 and then back down to $2.50 at the end of the year. In the next few years the price of gasoline will likely reach $5.00, and then retreat back down to $3.50. Then in subsequent cycles it will likely creep up to $6.00 and retreat back down to $5.00, up to $8.00, and back down to $6.00 and so on. In this scenario, the price will be creeping up little by little. People will eventually get used to the new higher price and life will go on as usual. Of course, this is the best scenario possible. Do you really believe that the United States will be that lucky?

Summing up, it is unlikely that the price will double every seven years exactly as the table above shows. In the past, the price went from $2 in 1972 to $35 in 1980, then all the way back down to $12 in 1999, and then up to $70 in 2005. The question is, "Is the price going to retreat some, before it goes to the next higher level, or is going to double two or three times before it retreats?" Looking in retrospect, over the next fifteen years or so, that question will be answered.

National Security

The case to eliminate imported oil does not just depend on predictions that the world will start running out of oil and that the supply of oil will get tight; there are other more important factors that have to be considered too. Perhaps a quote from Jimmy Carter during his presidency explains it best.

"Our decision about energy will test the character of the American people and the ability of the president and the Congress to govern. This difficult effort will be the "moral equivalent of war" – except that we will be uniting our efforts to build and not destroy." – Jimmy Carter, April 18, 1977.

Jimmy Carter deemed oil dependence a moral equivalent of war. During Jimmy Carter's presidency, the only real oil problem the United States faced was economical. Today terrorism, war, and a large number of massacred Americans are part of the overall challenge. The U.S. dependence on oil is no longer the moral equivalent of war; rather it is just a war, with all of its harsh realities. Had the U.S. adopted a policy not to import any oil from other countries fifty years ago, the oil production might have been booming today, without any threat of an oil war looming over this nation. Cheap foreign oil is what halted the U.S. oil production at home some fifty years ago. Had Americans not imported a single barrel of oil from abroad, a possible hypothetical scenario could have developed as follows:

At the point when the U.S. consumption caught up with U.S. production, the relationship between supply and demand would have determined the available oil reserves. When demand for oil exceeded the supply, the price of oil would have started to increase. The rising price of oil would have created opportunities to increase investment in exploration to find and produce more oil in the U.S. Higher gasoline prices would have made driving automobiles more expensive and oil usage would have decreased. At the same time, higher gasoline prices would have provided an opportunity to research alternative fuels such as biofuels, natural gas, hydrogen and electricity. It is hard to imagine this country being worse off today, had the United States adopted a path of better technology and stricter conservation, while absolutely rejecting oil dependency.

Rise of OPEC

OPEC is the only oil superpower in existence today. When OPEC talks, everybody listens. Where the U.S. needs Army superiority and relies on the most sophisticated weapons to maintain a superpower status, OPEC does not need anything. Just wave a magic wand and say, "the price of the oil needs to be higher," and it does. Now that is power.

The Organization of Petroleum Exporting Countries (OPEC) was created on September 17, 1960. The five original members were Iran, Iraq, Kuwait, Saudi Arabia and Venezuela. In addition to the five original founding members, Algeria, Indonesia, Libya, Nigeria, Qatar and the United Arab Emirates had joined OPEC by 1971. OPEC was formed in response to pressures of oil companies to reduce oil prices.

In 1959, the U.S. established a Mandatory Oil Import Quota, creating restrictions on the amount of foreign oil that could be imported to the United States. Reduced quantities of foreign oil coming to the U.S. has contributed to depressed prices of oil in the Middle East, where individual countries were too small to effect change. The primary purpose of OPEC was to engage in negotiations with other oil companies to obtain higher fair prices for crude oil. They also sought to establish a level of production, which produced a fair return on the investment to all participating countries. OPEC countries on their own had no power but together they gained leverage to deal in international oil markets dominated by the oil giants, the multinational oil companies.

In 1967, the Six Day War triggered the formation of The Organization of Arab Petroleum Exporting Countries

(OAPEC) and galvanized political pressures against the western countries supporting Israel. The Arab-Israeli conflict had transformed OPEC from its primary purpose of being able to obtain a fair price for their crude oil to an effective and determined biased political force. In 1973, after the Yom Kippur War, a conflict in which Egypt together with Syria attacked Israel, an enraged OPEC imposed an oil embargo against the U.S. The U.S. support and the U.S. supplied arms to Israel created tensions between the Arab World and the West, primarily the United States, and was the only reason for the embargo. Denied shipments of oil had created a crisis resulting in a fivefold oil price increase of the crude oil. The United States, for the first time, recognized the enormous power OPEC was able to exert, just by controlling the supply of oil. Politicians of both political parties pledged for this nation to become oil independent so this type of power display would not affect the United States again.

OPEC countries control about two-thirds of the world's total reserves and account for 40 percent of world oil production. Controlling such a large portion of the world's reserves has given OPEC a power, not just to control the production levels, but to control the price of oil as well. Largely due to the fact that OPEC members cannot get their act together, the U.S. experienced low oil prices during the 1980 to 2004 period. The production of oil of OPEC nations depends on the amount of estimated oil reserves these countries contain. The larger oil reserves the countries claim they have, the more oil they are allowed to produce, thus a system designed to limit production in order to keep the prices high. Limiting output quotas set the oil price. Overproduction and cheating of OPEC members in the past has created an excess of oil, keeping the price of crude oil low.

Due to the fact that Saudi Arabia is the largest oil producer in the Middle East, it has historically taken on the roll of cutting or increasing oil production as necessary, to keep the price of oil at a desired level. During the early 1980s, when the oil demand dropped dramatically, (caused by the Iran-Iraq war) Saudi Arabia's oil production had dwindled to one-third of what it produced just a few years before. By the mid-1980s, Saudi Arabia seized to provide the price control function and reestablished full oil production, hence flooding markets with oil. The decision of Saudi Arabia to flood the market was partly due to the refusal of other OPEC nations to reduce their production. As the price of oil dropped significantly, OPEC member nations had learned the power of excess capacity.

Today's OPEC admits that it has very little excess oil production capacity and it is being suggested that the lack of production capacity will significantly reduce OPEC's power to manipulate the price of oil. Nothing could be farther from the truth. Nothing can prevent OPEC nations from cutting oil production deliberately at any time they want to. Any such intentional cut would increase the price of crude oil. In fact OPEC, with this type of conduct, can control the price of oil at will. To end with, OPEC is free and fully capable to exercise their political power whenever they want to. And with the rest of the world pumping at full capacity, OPEC's pricing power and its influence is greatly intensified.

OPEC is not foolish, it knows that too high crude oil price will reduce the demand and provide incentives for consumers to seek oil alternatives much faster. OPEC will not create an environment in which the crude oil is no longer the cheapest fuel available. OPEC will test the price boundary before increasing the oil supply. Where exactly are these boundaries? European nations are comfortable paying $6.00 to $7.50 for a gallon of gasoline. This price reflects a large government gas tax; however, baring any tax on gas, $7.50 per gallon reflects a price of $250.00 per barrel of oil, a price which OPEC can easily conclude, using the European model, to be the next price target without any significant reduction in world demand.

Additionally, OPEC knows the real money will be made on the other side of the oil peak, when the production is decreasing. So for now, OPEC will keep supplying cheap oil, making the U.S. addiction to oil grow stronger. OPEC wants the rest of the nations, including the United States, to become a drug addict, with a cocaine-like addiction; thus "The Oil Hostage" of the oil cartel. OPEC does not want the world to worry and start thinking of oil alternatives seriously, not until it is too late. Once the oil supply starts to shrink, supply and demand will take over and the consumers (Americans) will begin to pay. While the U.S. is investing heavily in a new infrastructure, to make a changeover to an alternative source of energy, Americans will use a hefty portion of their hard earned paychecks to fill up their gasoline tanks.

The time to build and complete a new infrastructure is estimated to take between twenty-five and fifty years. Therefore, if the oil production is expected to begin to shrink in twenty-five years, then the time to set in motion the building of a new infrastructure is now. Americans need to understand the fact that OPEC could care less if the high price of oil puts an undue stress and pressure on world economies, particularly the U.S. economy. They need to understand that OPEC could care less if the United States' and the world's standard of living declines due to higher prices of oil, as long as the OPEC's bank accounts grow.

To summarize; OPEC is a group of nations which attempts to restrict the supply of oil to achieve a higher price, in effect eliminating competition. Since 1967, the U.S. has had to deal with unfriendly nations to get the necessary oil. The increased oil revenue enabled the Middle East nations, Iraq and Iran respectively, to obtain weapons to arm themselves, where they now possess the ability to disrupt oil flow, not just by reducing oil production but also by the use of force. To prevent oil flow disruptions, the United States has to get involved militarily. However, at present, the U.S. has to face more sophisticated armed forces, resulting in many more U.S. casualties. Despite these facts, the U.S. government continues on a path to larger oil dependency.

The free trade proponents, ignoring and dismissing the tens of thousands of casualties, continue to insist that dealing with OPEC is "free trade," as free trade was designed to be, and is the best possible course for this nation. It looks as though this is a free trade, in which OPEC is free to do as it pleases and the rest of the world must keep their mouths shut. The next section will examine free trade.

Free Trade vs. Protectionism

This brings the discussion to the everlasting question. Is oil a commodity that deserves free trade status or is oil a commodity that requires some kind of protectionist measures?

Free trade has been the foundation of America's economy ever since the country was founded. Discussing free trade on a global scale would mean talking about other countries removing trade barriers. The major trade barriers are tariffs, subsidies and quotas. Other examples of trade barriers are licenses, regulations, laws and bans on foreign products.

Free trade is when goods between countries are traded without any trade barriers.

Protectionism is a policy protecting nation's products from foreign competition. This policy uses high tax tariffs and quota restrictions.

Tariffs are high taxes that are added to imported goods to make these goods more expensive. The imported goods become less competitive with our country's goods and therefore less attractive to our consumers.

Subsidies are when the government pays their domestic producers, (financially supplements the domestic producer's goods), so these producers can sell their goods at a lower price than the foreign competitors.

Import quotas are limits set on the quantities of goods that can be imported to a country. (Restricting the amount of foreign cars coming to this country, within one year, is designed to protect the domestic automotive industry). The effect of allowing limited quantities of product into a country creates a shortage of otherwise plentiful goods.

These shortages drive up prices that consumers pay, thus making the imported products less competitive with the domestically produced goods.

Tariffs, subsidies and quotas are designed to give competitive advantage to domestic producers over the foreign competition. Each country uses these tools in one degree or another. Trade wars develop between countries when countries put up excessive barriers. For instance, the first country puts up barriers against goods of the second country and the second country is likely to retaliate and put up barriers against the goods of the first country, and the overall trade comes to halt. There are economists on both sides of this argument. Both passionately argue that their way is better for the U.S. Lets examine both points of view.

Free Trade

Free trade proponents argue that when a consumer is given a choice, he will usually buy the cheapest product available to him, meaning that the consumer is best served when there is free trade without any barriers at all.

- Free trade proponents will argue that imposing trade barriers causes higher prices, because the consumer has to pay not only the cost of the product but also the cost of the tariff.
- They argue that imposing trade barriers causes higher taxes, because not only does the consumer have to pay the cost of the tariff, but he also has to pay the cost of bureaucracies to enforce the tariffs.

- They argue that the net effect of trade barriers is the loss of jobs. They acknowledge that there are few special interest jobs saved by tariffs, but when a consumer has to pay higher prices for some products then they are left with less disposable income to purchase other products, which in turn causes loss of jobs in the other industries.
- They argue that groups that benefit imposing trade barriers are usually special interest groups like large corporations and unions, who through powerful lobbying can influence the politicians.
- They argue that it is unfair for a person earning $7.00 per hour to subsidize jobs earning $25.00 per hour.
- They argue that free trade will force countries to specialize in production of products where they have competitive advantage for their exports, and import products from other countries, which they cannot make as cheaply.
- They argue that the net result of protectionism is lower living standards and higher levels of bankruptcies.
- They argue that 'everyone' asking for tariffs or subsidies just wants to charge higher prices or earn higher wages without ever having to compete.

Of course free trade proponents are able to supply all the literature and all the numbers to substantiate their arguments.

Protectionism

On the other hand, the proponents of protectionism argue pretty much the opposite.

- They argue that free trade is destroying America's manufacturing industry. They point out that some industries have disappeared altogether. The U.S. textile industry is all but gone, the shoe industry is gone, and there are no TVs or many other electronics made in the U.S. today. Even the cornerstone industries, like the steel industry and the automotive industry, are being threatened and facing extinction.
- They argue that today's manufacturing base is about the same as it was in 1960. It has not kept up the pace with the work force increase.
- They argue that competing with countries with much cheaper labor costs forces domestic companies into bankruptcy. They argue that the U.S. is trading with regimes using slave or child labor, to gain competitive advantage.
- They argue that trade deficits cost Americans jobs.
- They argue that free trade is causing the United States to lose its technological advantage.
- They argue that American labor costs have to be protected to keep up existing standards of living or we will start going backward to resemble the countries we compete with.

For every argument one side brings forth, the other side comes up with a counter-argument of its own and it is usually pretty much the opposite point of view.

Both sides provide a passionate argument to establish the superiority of their position, in which each side has valid points. There certainly have to be some protectionist measures imposed, when the U.S. national security or the U.S. national welfare is involved, and again just imposing protectionist measures without a well-thought-of plan can be counterproductive. Take the domestic steel industry as an example.

In the 1970s, the U.S. domestic steel industry was dying out. Workers were being laid off because the steel industry was unable to compete with foreign steel. Poor protectionist measures were put in place and the industry was asked to modernize, to compete with foreign competition in the future. The ugly unfortunate byproduct of no competition was that the wages nearly doubled in the next ten years. The companies and the unions took advantage of the situation and used the taxpayer's money to better themselves. As that took place, the United States industry was forced to buy U.S. steel and had to pay up to 25 percent more than they would have to pay for the foreign products had no protectionist measures taken place. In the rest of the industry where the competition continued, the wage increases were considerably less.

The domestic automobile industry has similarly taken advantage of the tariff situation. As the tariffs were put in place on the foreign automobiles to enable U.S. automobile companies to compete with their foreign counterparts, the U.S. automakers increased the prices of automobiles at a much faster rate.

Oil - A Commodity

Oil is just as any other commodity. As a commodity it trades just as clothing, produce, electronics or automobiles. Oil is subject to the same rules of free trade. Tariffs could be placed on imported oil, just as easily as placing tariffs on wood coming from Canada. Large subsidies could easily be provided to the oil companies so they can compete with the Middle Eastern two-dollar oil, and quotas could be established to limit the amount of foreign oil coming to this country. So why has the oil industry been allowed to get in such a dire predicament, so deeply in trouble?

Oil is the most visible commodity to the consumers in this country. If one applies tariff to Canadian wood, for example, the consumer will not see a drastic price increase when buying homes or furniture. The large price increase of wood will be absorbed by the prices of the other products and labor costs. The price increase of homes will be marginal compared to the increase of wood. The cost of labor, including all other materials, will remain the same and even if the wood doubled in price, the price of a home may increase just by a few percent. Additionally, only people buying news homes would notice these price hikes and being relatively small in numbers these price hikes would be ignored.

If on the other hand a 100 percent oil tariff was imposed against foreign oil, the price of gasoline would nearly double and everyone would notice. A politician would certainly become uneasy trying to explain that type of tariff. In effect, foreign oil has been allowed to come to this country free of any tariffs for decades. Up to now, no one has even suggested any meaningful tariff, and if they did it was insignificant in the name only, not having any real lasting effect.

Giving large subsidies to the domestic oil industry has met with the same fate of exception. Why should the oil industry get subsidies, when they are making such large profits? But when domestically produced oil costs the U.S. oil companies $20 per barrel to produce, as compared with Middle East oil, which costs $2 per barrel, it can be easily recognized why the domestic oil drilling industries are disappearing. Quotas, which raise the price of an imported good by making it rare, and therefore more desirable, are just as unpopular. Again everyone is going to notice when they are applied to oil. One can see why politicians avoid using quotas on imported oil too.

Because of oil's visibility, instead of being treated as any other good that is being imported to this country, oil has developed trading parameters of its own. When it comes to oil, the proponents of free trade won big time. The world's supply and demand has dictated the price of oil without any interference of trade barriers whatsoever and politicians have developed a new parameter called "conservation" to replace the oil trade barriers.

Proponents of free trade and proponents of protectionism may argue if it was or wasn't a good idea to apply protectionist measures to the U.S. textile industry, the U.S. steel industry, or the U.S. automobile industry. They may argue if the customer was better served when the protectionist measures were applied, or if the customer would have been better served had they not been applied. They may argue whether jobs were lost or jobs were gained. In the industries in which protectionist measures were applied, the arguments of pro and against will go on forever.

When it comes to oil, everything that happened in this industry, be it good or bad, is the direct result of free trade since no real protectionist measures were ever applied. All the successes and all the failures associated with oil are the result of free trade left unchecked. The successes as well as the failures are numerous.

The successes are:

- The cheap price our customers paid for oil until recently was certainly a free trade success.
- It may be argued that without cheap oil this country would not be as prosperous as it is today; and as mentioned earlier, one out of every six jobs is related to the automotive industry in one way or another and the proponents of free trade may argue that a lot of jobs were created, and Americans are more prosperous then they would otherwise have been.

On the other hand, the failures are just as numerous:

- The cheap oil caused Americans to splurge big time, wasting oil. This waste made Americans dependent on imported oil, where close to 70 percent of the oil supply is now imported, letting other countries control the oil supply and the oil price.

- United States oil production has been declining for the past thirty years, causing loss of jobs in the oil industry.
- Americans have endured an oil crisis, oil shortages and wide price fluctuations in the last thirty years.
- American people are suffering bankruptcies because of the wide oil price fluctuations and the wide price fluctuations are common free trade characteristics.
- The U.S. is using its military strength to guard oil-rich regions to prevent oil disruption to its peoples; additionally, the U.S. interferes militarily to resolve other countries conflicts, risking American lives. And by doing that, the U.S. is losing young American soldiers in the process, over 22,000 casualties of war and counting.

Of course, one may ask if the United States has helped itself as a country by using "free trade," or would they be better off using some kind of protectionist measures. You be the judge.

Perfect World and Free Trade

When a consumer is given a choice he will usually buy the cheapest product available to him, meaning the consumer is best served when there is free trade, without any barriers. Society may specialize in the production of just a few products, increasing output and productivity while minimizing the cost of a specific product. Thus a country can be importing many goods that can be made cheaper in other countries, while exporting just a few goods it has specialized in. These are interesting points; they have a lot of merit and are worth a little more attention.

There has to be environment in which this type of thinking is viable. First, the type of product needs to be examined. Some products may be better candidates than others. Society might be more inclined to have free trade without any government interference, thus risking becoming totally dependent on other countries, if they traded with some innocent products such as clothing, cosmetics or electronics. Interruption of these products may increase prices but it is unlikely that the lifestyle of a society would somehow be altered. It is unlikely that the society would be in a state of panic. One can always wear old clothes, stop wearing cosmetics, or go to a theatre if a TV is not available.

On the other hand, when other products such as agricultural products, meats or medicines are involved, then the interruption of this supply could be detrimental to the society. Just the thought of a shortage of food or not having the medicines for children or the elderly has to make most people uncomfortable. To become dependent on other counties for the necessities of life is not to the best interest of a society. The U.S. certainly should not allow other countries with a much cheaper labor force to mass-produce agricultural goods, and to use free trade to undercut prices of American farmers and drive them out of business. Large firms could undercut prices of their products and drive competition into bankruptcy. This type of predatory pricing practice is called dumping and is considered anti-competitive and illegal in the United

States. Unfortunately, if discovered late, it may certainly cause a lot of unnecessary difficulties.

The Japanese government, for example, is subsidizing their Japanese farmers producing rice. The Japanese pay five times as much for their local rice than they would if they had bought it on the open market from other countries. They do that to preserve their farming industry, the domestic production. European countries are doing precisely the same. They too pay heavy taxes for their domestic farm subsidies. It sure seems that food is off the table in many of the countries where free trade is being applied.

The second point to examine is whom the U.S. trades with. Friends can become enemies as regimes change. Wars interrupt imports that society grew to depend on. Political views change and products that would have been otherwise marked for export remain frozen. Certainly, there are a lot of world events out there that could interrupt the supplies. Interruption of any product would be a temporary situation only. Other countries would fill the gap eventually. In the meantime, the suffering will be directly related to what product is involved.

Flu vaccine is a good example. Because the U.S. became totally dependent on England, the only supplier for imports of the flu vaccine, there was no flu vaccine available for most of 2004, when England's supply of flu vaccine became contaminated. Eventually the U.S. was able to obtain new quantities from other countries, but not without real shortages and widespread panic.

The United States business model is unlike any other regime in the world. The U.S. government provides favorable environment for American companies, so these companies can prosper. The companies in turn use their ingenuity to provide Americans with their products. Too many regulations and businesses go bankrupt; too few regulations and some companies will take advantage of the situation. What an awesome power, that of the U.S. government.

Protectionist measures are many times applied at the wrong places with the wrong products. Who cares where the radios, batteries and steel come from? Who cares where the automobiles are made because someone, somewhere, is going to make these products, and if not, the USA will figure it out and make these products on their own. If Europe refuses to make cars then the Asian countries will, and if Japan refuses to supply the radios then Europe will make up the slack.

With oil however, it is much different. If the Middle East stopped producing oil, then there would be nowhere else to go, and that puts the petroleum in a category by itself. So the question is: Does oil fall in the category of cosmetics and clothing, where no one would miss it, or does oil fall in the category of food and medicines which Americans could not do without? Perhaps applying this to a disaster situation, such as a hurricane, will help you come up with the answer.

U.S. Oil Companies – OPEC Connection

Oil companies are in business to first produce and sell their oil, and second to act as a broker to bring oil from other parts of the world, refine it, and then sell it at the profit, without which these companies would go out of business. Any mention of electric power, wind power, solar panels or nuclear power is a nonstarter for these oil giants. It is easy to recognize why the U.S. oil industry is out there acting as if there was not a worry in the world. There is no problem, so don't scare the public. U.S. oil companies provide optimistic information to Americans. One spokesman for a U.S. oil company has just recently said: "We are looking at super spikes in oil prices in the next five to six years and then prices will come back to normal." What exactly are these super spikes?

No matter how profitable the oil business is, once an alternative fuel begins to replace the oil, these oil giants will begin to downsize. Who is kidding whom, they know it too. They know that the time is closing in, when oil will be replaced by another fuel. It is no wonder that the exploration is slowing and that there are so many mergers between the existing oil companies. The large and strong giants are gobbling up the small oil outfits; the strong will survive and the small will disappear.

A noticeable puzzling problem arises when one examines the oil companies' relationship with OPEC. Oil companies portray themselves as customer-friendly problem solvers. They claim that they want to provide plenty of oil to the U.S. customer and that they want to satisfy the customer's demand at the best possible price. Most Americans believe that the oil giants such as BP, Exxon, Amoco and Sunoco are looking out for American interest by providing Americans with a competitive oil price. Most Americans believe that these oil producers compete with other countries, including OPEC, and can influence the price of world oil, thus able to provide Americans with a somewhat stable oil price.

Oil War Games

The truth is very much the opposite. The U.S. oil companies along with many other countries of the world produce oil at their maximum capacity, and because they do not have any additional production reserve capacity, none of these countries have any control over oil prices. They do not compete with OPEC to establish a competitive oil price; rather they rely on OPEC to set a price, which OPEC deems appropriate. Thus, OPEC sets the price and the U.S. oil companies just go for the ride.

Once this relationship of price control is realized, one may conclude that, although the U.S. oil companies are not direct OPEC members, they for all intents and purposes belong to OPEC nevertheless. Because of this non-competitive relationship with OPEC, it suggests a potential conflict of interest. The U.S. oil companies produce seven million barrels of oil per day domestically, so when OPEC forces the price of oil to increase by, say, $10 per barrel by manipulating the oil supply, the U.S. oil producers earn $70 million per day profit on their seven million barrels without even raising an eyebrow. To put it in different terms, for every $10 increase forced by OPEC nations, U.S. oil giant's profits increase by $25.5 billion per year. This is why the business community has seen such huge oil profits reported by the U.S. oil producers from 2003 to 2006.

So a valid question is; what do the big U.S. oil giants really want? Are they promoting policies that are in the best interest of the United States or are they promoting and supporting the OPEC agenda, to earn more money without doing anything? They will need to make their true agenda known. Rather than talking about the plentiful oil supplies and the oil prices dropping in the near future, they need to provide the American consumers with some real explanations. Otherwise this newly found wealth of the U.S. oil companies will, and should, be viewed by Americans and the U.S. government as a payment, a bribe, a reward compensation to the U.S. domestic oil giants by OPEC, for the U.S. oil companies to keep silent and promoting U.S. oil dependency so OPEC can make as much money as physically possible.

For more detailed information go to,

http://en.wikipedia.org/wiki/European_Union
http://en.wikipedia.org/wiki/World_trade_Organization
http://en.wikipedia.org/wiki/NAFTA
http://en.wikipedia.org/wiki/OPEC
http://en.wikipedia.org/wiki/Gulf_War
http://en.wikipedia.org/wiki/Iraq_War
http://en.wikipedia.org/Iran-Iraq_War

3.
Caught off Guard

United States was certainly caught off guard. As the American people enjoyed the cheap oil party and saw the U.S. oil production peak, the U.S government should have seen the red flag. The U.S. government should have seen the red flag in 1959, when America began to use more oil than they could produce. Maybe, the U.S. government should have seen the red flag during the 1973 crisis, when Americans stood in long gasoline lines. And maybe the current administration of 2006 should see the red flag now and finally realize that something needs to be done.

During World War I, the U.S. exported two-thirds of the world's oil supply. The United States along with countries in Europe have provided the world with plenty of oil, at a price of about $1 per barrel. World War II forced the U.S. to begin to import oil, and in the early 1950s the U.S. began to import oil from the Middle East. Since then, the dependence on foreign oil has grown from about 9 percent of the total domestic consumption in 1960 to 65 percent in 2005. This ratio is expected to increase, where by year 2025 the ratio could reach 80 percent if the current policy is allowed to continue.

Is the U.S. government at all serious about the energy situation or are the politicians currently just using the oil situation to score points with the public to secure votes? If one listens to the government officials today, it would certainly point to the latter being true. It would seem as if the people in the government have no clue as to what is going on. If someone announced the oil shortage on the floor of the house, all of them would act as if they were struck by the headlights of an oncoming car; as if they heard the news for the first time. The fact is that since the 1973 Oil embargo, every president, every senator, and every congressman has been wholly aware of the energy situation. How is it never mentioned?

There is the government's will, a desire to do anything. A perfect example of the U.S. government at work is the discussion of a windfall tax on the oil companies in October and November of 2005. Due to high gasoline prices in 2005, the oil companies made substantial profits. As prices of gasoline at the pump rose to over $3 per gallon, people became jumpy and the politicians noticed. With bipartisan support to enact the windfall tax, the lawmakers came up with eleven different proposals to basically punish the oil companies for making a profit. One of these many proposals was to enact a 50 percent tax on all the oil profits above $40. As the gasoline price receded a little, to $2.50 per gallon, bipartisanship fell apart and the appetite to do something had dissipated. By Christmas of 2005 the windfall profit discussion was dead.

No matter how dumb the idea was, at least it got the politicians talking. The sad part, however, is that this approach would have not solved anything. The price at the pump for American consumers would remain the same. The same gasoline demand and the same lack of gasoline supply would continue to exist. None of these concerns were addressed. The only thing that would get accomplished by taxing the oil companies would be that some money would change hands. Large sums of money would be taken from the oil companies and would find its way to the government's coffers. In the government's hands without any commitment to solve the energy problem, the money would just be used to pay for some selected government programs.

Rather than taking the money from the oil companies in the form of large oil taxes, a better use of the government's time would have been perhaps for the government to tell the oil companies that they have to use this money, the newly obtained riches that were caused by the persistence of OPEC to raise the price of oil, for research and development. Direct these oil giants to use this money for oil exploration and research of energy alternatives. Additionally tell these companies that if they fail to do what the U.S. government asks of them, the money will be taken away and given to a new entity created just for that specific purpose. Let's examine how various administrations in the past dealt with the oil issue problems.

President Dwight Eisenhower

The threat to national security was recognized as early as 1957, when the imports were relatively small. In 1957, recognizing that oil imports may threaten national security; President Eisenhower crafted a voluntary program restricting oil imports. Voluntary quotas were adopted, relying on the cooperation of the oil industry. Because the voluntary quotas had proven ineffective, President Eisenhower instituted a Mandatory Oil Import Program, establishing mandatory quotas. The program stipulated that imports would not exceed 9 percent of the total demand. Unfortunately the program failed. Over the next several years lobbyist groups managed to make many exceptions to the rule. The oil companies managed to get what they wanted and the program seized to do its function. This system remained in effect until 1974, when a new fee system was established by President Nixon, providing for unrestricted access to oil imports.

President Richard Nixon / President Gerald Ford

President Nixon was faced with the first oil energy crisis in U.S. history when the organization of Arab Petroleum Exporting Countries declared an oil embargo against the United States. A fivefold price increase of oil had caused shortages, high prices, long lines in the gasoline stations and frustrated many Americans. On November 7, 1973, a path toward Energy Independence had been reborn. President Nixon launches "Project Independence." Project

Independence was designed to make the United States self-sufficient in energy by 1980. President Nixon said, "Our ability to meet our own energy needs is directly linked to our continued ability to act decisively and independently at home and abroad in the service of peace; not only for America, but for all nations in the world." President Nixon challenges the technology to help free the United States from dependence on imported oil.

After the resignation of President Nixon; President Ford continued the plight and on December 22, 1975 he signed the Energy Policy and Conservation Act (EPCA). This regulation was to establish strategic oil reserves of 1 million barrels, to increase the existing energy production, and to develop new energy supplies including nuclear, coal, solar and geothermal. President Ford had launched a voluntary automobile efficiency program to increase gasoline fuel efficiency. His effort led to Congress passing the Energy Policy Conservation Act on December 22, 1975, which was supposed to double the fuel efficiency of the U.S. automotive fleet in ten years. (More on this in the conservation chapter.)

President Jimmy Carter

If Americans have any doubt that the U.S. government knew of the upcoming oil crisis for more than thirty years, then they should read President Carter's televised speech given on April 18, 1977. (Because this speech addresses energy problems only, it is reprinted at the end of this chapter in its entirety.) It is almost frightening to see to what extent President Carter's administration understood the situation developing in front of them. You will be amazed with the accurate prediction of the U.S. energy situation, which came to pass the day after 9/11. To this date (1957 to 2006), President Carter is the only president who stood in front of the whole nation and told the truth on the subject of energy, without regard to how it would affect him politically.

- President Carter understood how much oil we had wasted in the '40s, '50s, and '60s.
- He understood that the United States oil production was already dropping.
- He understood that at the rate the U.S. oil imports were increasing, the U.S. could see the world peak of oil production, (the world demanding more oil then it could produce), within a decade.
- He understood the economic disaster the oil shortages could bring to the U.S. economy. As he put it, the "do-nothing policy could lead to a national catastrophe."
- He understood that the nations' independence was being threatened.

He adopted a ten-year plan, a path toward reducing the oil consumption using large-scale conservation. His plan included increasing automobile fuel efficiency, insulating homes and businesses, and finally, to replace a large part of the oil consumption with coal and solar energy. President Carter forecasted that upcoming oil problems would be facing the United States by 1985 if nothing were done.

In 1979, Carter's presidency had been handed yet another energy crisis. Early in 1979, the Iranian revolution began, where the Shah of Iran fled the country and Ayatollah Khomeini rose to power. The Iran oil production dwindled to nothing and the world began to consume 2 million barrels of oil per day more, than it could produce. President Carter's office administered a price control, which resulted in shortages and long lines at the gasoline stations, just as in the 1973 situation. Rationing was the popular talk in Congress. Responding to the energy shortages in the summer of 1979, President Carter had taken steps to decontrol the oil prices. He even went as far as to establish temperature restrictions in nonresidential buildings.

On June 20, 1979, President Carter proposed a program to increase the use of solar energy, and on June 30, 1980, President Carter signed the Energy Security Act, which included the Synthetic Fuels Corporation Act, the Biomass Energy and Alcohol Fuels Act, the Renewable Energy Resources Act, the Solar Energy and Energy Conservation Act, the Solar Energy and Energy Conservation Bank Act, the Geothermal Energy Act and the Ocean Thermal Energy Conversion Act. On July 15, 1979, President Carter proposed an $88 billion effort to produce oil from coal and shale, thus launching the "Oil Shale" business.

As if the two crises for one term of a presidency were not enough, President Carter faced yet another problem. An accident at Three Mile Island on March 28, 1979, sealed the fate of the new nuclear power plants. Since then, no new nuclear power plants were started and the

U.S. has plunged into the dark ages as it pertains to nuclear research and development.

Maybe President Carter was plagued by bad luck and maybe it was good that he did not get reelected. Nevertheless, Americans should be thankful to this president for having the guts to come forth and tell them "the whole truth and nothing but the truth about imported oil," especially when the truth included such unpopular solutions. The American public unfortunately did not want to hear that. Whatever happens during a president's term is the president's fault; at least that is how the story goes, and so Americans concluded that enough is enough and a new president, a "White Knight" of sort, needed to be elected to rescue Americans from the "Big Bad Wolf," the horrors of imported oil.

President Ronald Reagan

When President Carter's term ended so did his policies. President Ronald Reagan pretty much scraped Carter's proposals the first day he got into office. On January 28, 1981, President Reagan signed executive order, 12287 to decontrol crude oil and refined petroleum products. This does not mean that he was a bad president, not by any means; it just means that his approach to the energy solution was completely different. President Reagan was from the school that would let the forces of supply and demand fight it out. His idea was not of conservation but to get more oil to satisfy American needs, and he was a master at that.

As he was trying to outspend the Soviet Union into bankruptcy he began to be known as tough, not willing to give into the forces of evil. The U.S. business partners in the Middle East knew President Reagan's reputation, especially Saudi Arabia. The thought of the United States not being there if and when Saudi Arabia needed them, especially during the Iran-Iraq war which lasted from September 22, 1980 to August 20, 1988, was enough for Saudi Arabia to increase the oil production when Ronald Reagan courteously and graciously asked them to. As the oil prices dropped and oil became plentiful, Americans became content and happy again.

President George Bush Sr.

President George Bush had continued with Reagan's policies. He had enjoyed a relatively stable price of oil so the oil problems of the future were not a priority. Let supply and demand fight it out. He too, however, faced a war. The Iraq invasion of Kuwait began on August 2, 1990. Economic sanctions by the United Nations were implemented immediately and a coalition of forces of some thirty nations began to drive the Iraqi forces out of Kuwait in January of 1991. Coalition forces met with minimal casualties as the main battles were conducted using U.S. Air force superiority.

On February 20, 1991, following the war, President George Bush launched his National Energy Strategy. His strategy was to promote energy conservation and increase energy supplies. This strategy was not designed to lead the U.S. toward oil independence, rather just to slow down the unprecedented rise of consumption. On October 24, 1992, President Bush signed the Energy Policy Act of 1992, implementing his National Energy Strategy.

The proposed policies, summarized, are:

- Diversify the U. S. sources of energy supplies.
- Increase energy efficiency in the entire industry.
- Develop technologies to fuel-efficient engines.
- Begin the use of electric vehicles.
- Scrap the old gas-guzzlers.
- Increase electricity efficiency.
- Reduce energy consumption in federal buildings by 20 percent by the year 2000.
- Reduce dependence on foreign oil. Diversify oil imports.
- Increase the U.S. oil production.
- Drill for oil and gas in ANWR.
- Speed up offshore drilling.
- Develop recovery technologies to recover our existing supplies.
- Increase the use of natural gas and coal.
- Increase the production and use of renewable energy.
- Develop and increase the use of wind energy.
- Develop and increase the use of solar energy.

Just as President Carter before him, President Bush had come up with an ambitious energy plan; again the plan did not get anywhere. Americans will never know whether or not his plan would have worked since they had decided for these policies not to continue.

President Bill Clinton

Like President Bush, President Bill Clinton had enjoyed stable oil prices and therefore was able to utilize President Ronald Reagan's strategy; do nothing and let the supply and demand resolve any problems. Plentiful and cheap oil suggested that no conservation was needed. Americans were satisfied and so neither President Clinton nor the U.S. Congress decided to rock the boat. However, he too will not remain without an energy legacy. In September 1993, President Clinton's administration formed "A Partnership for a New Generation of Vehicles." A pact between President Clinton and the three biggest automakers, specifically Ford, Chrysler and General Motors, was formed. The specific purpose was to develop a vehicle, within the next ten years, with a fuel economy of 80 miles per gallon, without sacrificing performance, cost and safety. Upward of $1.25 billion had been spent, yet the Detroit big three had produced nothing. The President Clinton energy legacy will likely go down in history as follows:

The anti-Clinton group concludes that just prior to reelection a favorable partnership was established between the Clinton administration and Detroit, so the Clinton administration could benefit politically and at the same time Detroit could receive a large payback in the name of research. Many, including a government representative, have said that the partnership was formed to avoid the increase of the federally mandated Corporate Average Fuel Efficiency Standards, so Detroit could continue to produce large gas-guzzlers, the bread and butter of Detroit automakers.

The pro-Clinton group speculates that $1.25 was not enough and that more money was needed for research and development to produce success. They speculate that the Republicans, led by Gingrich, failed to appropriate adequate funds and therefore the failure of Detroit could have been expected.

The Detroit failure had, however, produced an indirect positive result for American consumers. While Detroit had been hoarding the government's money to provide votes for the Clinton administration, foreign automakers, especially Toyota and Honda, had been developing the high-efficiency automobile. In 1997, just four years after the Clinton-Detroit partnership had been announced, Toyota unveiled a high fuel efficiency electric-hybrid car. In 2000, it began to sell its Prius model for about $20,000. Honda similarly had begun to sell its high fuel efficiency Insight for about $19,000. As of 2006, both of these automakers were just a few miles short of the Clinton-Detroit 80 miles per gallon goal and both had been able to accomplish this without any government subsidy.

President George W. Bush

At the start of his term, President George W. Bush's solution was to continue on the same path as his predecessors. The events of 9/11 had changed everything. In many ways you may say that, when energy is of concern, because of what happened on 9/11, Ronald Reagan's era ended and Jimmy Carter's era reemerged. It is as if it was interrupted for twenty years and now is resuming its course. The tragedy of 9/11 had exposed, not just to the U.S. but also to the whole world, that if terrorists can blow up buildings within the U.S. borders and kill thousands of Americans, they are capable of disrupting the U.S. oil flow from anywhere in the world. President George W. Bush has to deal with a much more difficult predicament in the oil situation, than did President Carter. President Carter had predicted that long-lasting oil shortages and interruptions would develop some ten to twenty years into the future, without any protracted interruptions in between. After 9/11, President George W. Bush could realize sudden, drastic oil shortages over night, and so a need to proceed differently developed.

He had to assume the responsibility of making sure that today's oil supply continues to flow uninterrupted, thus the Middle East presence. Now, however, an end strategy or the second phase of a plan needs to develop. President Bush now needs to take a cue from the Carter administration, which is to begin a path toward energy independence. The government now has to realize that the U.S. is not going to be able to guard the oil supply forever, and therefore some form of a conservation policy including steps that will lead the U.S. away from imported oil has to be taken. Rather than establish a comprehensive policy to energy independence, President Bush continues in Clinton's footsteps. He continues to throw money at Detroit without requiring results.

In January 2002, President Bush replaced Clinton's Partnership for a New Generation of Vehicles with his own version, "A Freedom Car Initiative." This initiative has scrapped the 80 miles per gallon automobile goal with a new goal; the automobile powered by a fuel cell requiring hydrogen as a fuel. The program is being introduced to the American public as being able to end oil dependence and reduce green house emissions. President Bush's administration promised Detroit's big three, Ford, Chrysler, and General Motors, $1.2 billion for research; virtually mimicking what President Clinton had done just a few years earlier. Again Detroit was not required to produce anything.

On January 28, 2003, in his State of the Union, President Bush unveiled a second part to his program in which he asked Americans to join him to make the air cleaner and our country much less dependent on foreign sources of energy. Another $1.2 billion went to his "Hydrogen Fuel Initiative," with the FY 2004 budget of $159 million and the FY 2005 budget of $227 million. This research initiative was to find ways to lower the production cost of hydrogen, to develop hydrogen storage so automobiles could travel in excess of three hundred miles before refueling, and to create affordable hydrogen fuel cells. The timetable given by the research initiative was as follows: If the research was successful and if the fuel cell automobiles could begin to be sold to the public by 2020, then by 2040 the U.S. could significantly reduce oil imports. According to the Department of Energy, the development of fuel cell could reduce the U.S. oil demand by 11 million barrels per day by 2040. This projection sounded optimistic and Americans gave it the benefit of the doubt.

Reproduced is a part of the speech given by President Bush, dealing with hydrogen automobiles.

Caught Off Guard

A partial transcript of the President Bush State of the Union Address from Wednesday, January 29, 2003:

> "In this century, the greatest environmental progress will come about not through endless lawsuits or command-and-control regulations, but through technology and innovation.
>
> Tonight I'm proposing 1.2 billion dollars in research funding so that America can lead the world in developing clean, hydrogen-powered automobiles.
>
> A simple chemical reaction between hydrogen and oxygen generates energy, which can be used to power a car while producing only water, not exhaust fumes.
>
> With a new national commitment, our scientists and engineers will overcome obstacles to taking these cars from laboratory to showroom, so that the first car driven by a child born today could be powered by hydrogen and pollution-free.
>
> Join me in this important innovation to make our air significantly cleaner and our country much less dependent on foreign sources of energy."

Just three years later, in January 2006, in his State of the Union Address, hydrogen was out and more fuel was back in. The hydrogen program, in which billions of taxpayer dollars were being spent, received the attention of half a sentence. New proposals had been mentioned such as better batteries for electric cars and new research in renewable sources of energy, including nuclear power which was taboo for decades. Additionally, ethanol was proposed as a new fuel, capable of replacing some of the imported oil within six years. Finally, the last paragraph of his speech on energy deals with reducing oil imports.

> *"Breakthroughs on this and other new technologies will help us reach another great goal: to replace more than seventy-five percent of our oil imports from the Middle East by 2025. By applying the talent and technology of America, this country can dramatically improve our environment, move beyond a petroleum-based economy and make our dependence on Middle Eastern oil a thing of the past."*

This paragraph claims to reduce the oil imports from the Middle East by 75 percent. The problem is that the U.S. imports only about 2.5 million barrels of oil from the Middle East and 75 percent of the Middle East 2.5 million barrels amounts to less than 2 million barrels per day. As mentioned earlier, in 2003 the Department of Energy estimated a reduction of 11 million barrels per day by 2040, and that would make it 75 percent of the 14 million barrels per day the U.S. currently imports. The president's 2006 estimate is about 2 million barrels per day by 2025 and that is just unacceptable; especially when it is projected that the U.S. will need an additional 8 million barrels per day by the time 2025 arrives. Better results are possible and should be pursued. Who knows, maybe President Bush really meant 75 percent of all of the oil imports, rather than just the Middle East oil. Maybe the speechmakers inserted the words "Middle East" for special dramatic effect. In any case, the president and the government should clarify their claims and not leave everyone guessing.

The part of the speech that deals with energy is reprinted.

January 31, 2006
STATE OF THE UNION ADDRESS BY THE PRESIDENT
United States Capitol – Washington, D.C.

"Keeping America competitive requires affordable energy. And here we have a serious problem: America is addicted to oil, which is often imported from unstable parts of the world. The best way to break this addiction is through technology. Since 2001, we have spent nearly 10 billion dollars to develop cleaner, cheaper and more reliable alternative energy sources – and we are on the threshold of incredible advances.

So tonight, I announce the Advanced Energy Initiative – a twenty-two percent increase in clean-energy research at the Department of Energy – to push for breakthroughs in two vital areas. To change how we power our homes and offices, we will invest more in zero-emission coal-fired plants, revolutionary solar and wind technologies and clean, safe nuclear energy. (Applause.)

We must also change how we power our automobiles. We will increase our research in better batteries for hybrid and electric cars and in pollution-free cars that run on hydrogen. We'll also fund additional research in cutting-edge methods of producing ethanol; not just from corn, but from wood chips and stalks or switch grass. Our goal is to make this new kind of ethanol practical and competitive within six years. (Applause.)

Breakthroughs on this and other new technologies will help us reach another great goal: to replace more than 75 percent of our oil imports from the Middle East by 2025. (Applause.) By applying the talent and technology of America, this country can dramatically improve our environment, move beyond a petroleum-based economy and make our dependence on Middle Eastern oil a thing of the past." (Applause.)

If it is a matter of national security, if someone threatens this nation or the U.S. way of life or if someone threatens an economic collapse of this nation, then the government better come to the rescue of all Americans. In fact, the government is the only entity that can handle a problem of this magnitude. The government will get involved eventually. It may not be this president or the current congress and senate; it might not even be the government that comes next, but rest assured that when the gasoline prices reach levels at which Americans take notice, like $10 to $20 per gallon, the government will get involved.

Once Americans understand the awesome power of the U.S. government to enact legislation, they will begin to ask why so many of the problems have not been resolved. Imagine the U.S. Congress passing a law, telling the automotive industry to double the gasoline efficiency of the U.S. automobiles or face stiff penalties. Imagine

Congress being able to call executives of the big oil companies to argue in front of them as to why they should not be faced with a windfall tax. What an awesome power.

Rather than to fix these problems, congressmen and senators squabble about whether or not the oil companies are price gouging, as if somehow the oil companies are responsible for the high oil prices. Out of the 435 congressmen and 100 senators of both political parties, including the present administration, not a single one acknowledges that perhaps OPEC has something to do with the high oil prices Americans currently pay. None of them are currently willing to acknowledge that perhaps it is not the U.S. oil companies but OPEC, and maybe even the government, who is the true U.S. enemy when it comes to oil pricing.

Let's go through an example to see price gouging at work:

> One barrel (42 gallons) of imported oil costs $75.00 or $1.80 per gallon.
>
> Tax, which most states impose, is 50 cents per gallon on the average.
>
> The price of gasoline of $2.75 per gallon, on the average, leaves around 45 cents for the oil companies.

So, the OPEC countries charge $1.80 for a product that cost 5 cents to produce. States charge 50 cents for nothing. Oil companies receive 45 cents for all of their work; they refine the crude oil into gasoline, distribute it through a complex infrastructure of gasoline stations and sell it to American customers, while still managing to make profit. It does not take a genius to recognize where the price gouging is happening. Yet the lawmakers and the media continuously point fingers at the oil companies, accusing them of price gouging.

Want to see real price gouging at work? Just look at natural gas. Compared to oil, natural gas is produced almost entirely domestically. In the summer, natural gas costs anywhere between $4 and $6 per Mcf. In the winter, during heating season, natural gas costs approach $20, a 400 percent price increase within a three-month period. This phenomenon happens every year. Putting it little differently, in the summer the gas producers make acceptable profit by charging $5 and in the winter they make a killing charging $20. No additional work, just lots of additional profit.

Of course, as the argument goes, it is the law of supply and demand. That would be a valid point if that happened only during some extraordinary occasions, such as the U.S. has experienced an unexpected brutally cold winter or some natural catastrophe such as a hurricane or earthquake that had shut down a significant portion of natural gas production. In this case however, it seems that there is always enough of a supply of natural gas at the high winter prices. One would think that the drastic price swings have happened so many times that the typical average demand of the upcoming season could be properly anticipated and the supply of natural gas could perhaps be increased so it would become plentiful at the $5 summer price throughout the entire year. But it seems that the gas suppliers project the new winter demand as follows: How much natural gas is going to be needed at the $20 price level during the typical average season? That is an accurate definition of price gouging.

Either there is a real natural gas shortage in the United States or there is an unprecedented scheme of deliberate price gouging going on. In any case, it should be investigated. If indeed the U.S. is experiencing the large shortages of natural gas as the winter price suggests, then Americans should be told. The lawmakers should not repeatedly tell the consumers that the U.S. has an adequate supply of natural gas. If, on the other hand, the U.S. has an ample supply of natural gas, then the winter supply should be correctly predicted and the true supply should be produced to keep the price of natural gas the same throughout the year. The winter price exploitation should stop.

None of the lawmakers or the media ever questions any of this. To all of them this phenomenon seems perfectly normal. The strangest thing of all is that there are no unfriendly OPEC nations involved with natural gas; only U.S. companies and the U.S. government. The monopoly of the natural gas suppliers provides the same unfriendly service OPEC does with their oil. It sure seems that it is more popular to blame large oil companies and gasoline, rather than OPEC or natural gas suppliers.

The President's Proposed Energy Policy*
(reproduced in its entirety)
Jimmy Carter
April 18, 1977

Tonight I want to have an unpleasant talk with you about a problem unprecedented in our history. With the exception of preventing war, this is the greatest challenge our country will face during our lifetimes. The energy crisis has not yet overwhelmed us, but it will if we do not act quickly.

It is a problem we will not solve in the next few years and it is likely to get progressively worse through the rest of this century.

We must not be selfish or timid if we hope to have a decent world for our children and grandchildren.

We simply must balance our demand for energy with our rapidly shrinking resources. By acting now, we can control our future instead of letting the future control us.

Two days from now, I will present my energy proposals to the Congress. Its members will be my partners and they have already given me a great deal of valuable advice. Many of these proposals will be unpopular. Some will cause you to put up with inconveniences and to make sacrifices.

The most important thing about these proposals is that the alternative may be a national catastrophe. Further delay can affect our strength and our power as a nation.

Our decision about energy will test the character of the American people and the ability of the president and the Congress to govern. This difficult effort will be the "moral equivalent of war" – except that we will be uniting our efforts to build and not destroy.

I know that some of you may doubt that we face real energy shortages. The 1973 gasoline lines are gone and our homes are warm again. But our energy problem is worse tonight than it was in 1973 or a few weeks ago in the dead of winter. It is worse because more waste has occurred and more time has passed by without our planning for the future. And it will get worse every day until we act.

The oil and natural gas we rely on for 75 percent of our energy are running out. In spite of increased effort, domestic production has been dropping steadily at about six percent a year. Imports have doubled in the last five years. Our nation's independence of economic and political action is becoming increasingly constrained. Unless profound changes are made to lower oil consumption, we now believe that early in the 1980s the world will be demanding more oil that it can produce.

The world now uses about 60 million barrels of oil a day and demand increases each year about 5 percent. This means that just to stay even we need the production of a new Texas every year, an Alaskan North Slope every nine months, or a new Saudi Arabia every three years. Obviously, this cannot continue.

We must look back in history to understand our energy problem. Twice in the last several hundred years there has been a transition in the way people use energy.

The first was about two hundred years ago, away from wood – which had provided about 90 percent of all fuel – to coal, which was more efficient. This change became the basis of the Industrial Revolution.

The second change took place in this century, with the growing use of oil and natural gas. They were more convenient and cheaper than coal, and the supply seemed to be almost without limit. They made possible the age of automobile and airplane travel. Nearly everyone who is alive today grew up during this age and we have never known anything different.

Because we are now running out of gas and oil, we must prepare quickly for a third change, to strict conservation and to the use of coal and permanent renewable energy sources like solar power.

The world has not prepared for the future. During the 1950s, people used twice as much oil as during the 1940s. During the 1960s, we used twice as much as during the 1950s. And in each of those decades, more oil was consumed than in all of mankind's previous history.

World consumption of oil is still going up. If it were possible to keep it rising during the 1970s and 1980s by 5 percent a year as it has in the past, we could use up all the proven reserves of oil in the entire world by the end of the next decade.

I know that many of you have suspected that some supplies of oil and gas are being withheld. You may be right, but suspicions about oil companies cannot change the fact that we are running out of petroleum.

All of us have heard about the large oil fields on Alaska's North Slope. In a few years when the North Slope is producing fully, its total output will be just about equal to two years increase in our nation's energy demand.

Each new inventory of world oil reserves has been more disturbing than the last. World oil production can probably keep going up for another six or eight years. But some time in the 1980s it can't go up much more. Demand will overtake production. We have no choice about that.

But we do have a choice about how we will spend the next few years. Each American uses the energy equivalent of 60 barrels of oil per person each year. Ours is the most wasteful nation on earth. We waste more energy than we import. With about the same standard of living, we use twice as much energy per person as do other countries like Germany, Japan and Sweden.

One choice is to continue doing what we have been doing before. We can drift along for a few more years.

Our consumption of oil would keep going up every year. Our cars would continue to be too large and inefficient. Three-quarters of them would continue to carry only one person – the driver – while our public transportation system continues to decline. We can delay insulating our houses and they will continue to lose about 50 percent of their heat in waste.

We can continue using scarce oil and natural gas to generate electricity, and continue wasting two-thirds of their fuel value in the process.

If we do not act, then by 1985 we will be using 33 percent more energy than we do today.

We can't substantially increase our domestic production, so we would need to import twice as much oil as we do now. Supplies will be uncertain. The cost will keep going up. Six years ago, we paid $3.7 billion for imported oil. Last year we spent $37 billion —nearly ten times as much— and this year we may spend over $45 billion.

Unless we act, we will spend more than $550 billion for imported oil by 1985 – more than $2,500 a year for every man, woman and child in America. Along with that money we will continue losing American jobs and becoming increasingly more vulnerable to supply interruptions.

Now we have a choice. But if we wait, we will live in fear of embargoes. We could endanger our freedom as a sovereign nation to act in foreign affairs. Within ten years we would not be able to import enough oil, from any country, at any acceptable price.

If we wait and do not act, then our factories will not be able to keep our people on the job with reduced supplies of fuel. Too few of our utilities will have switched to coal, our most abundant energy source.

We will not be ready to keep our transportation system running with smaller, more efficient cars and a better network of buses, trains and public transportation.

We will feel mounting pressure to plunder the environment. We will have a crash program to build more nuclear plants, strip-mine and burn more coal, and drill more offshore wells than we will need if we begin to conserve now. Inflation will soar, production will go down, people will lose their jobs. Intense competition will build up among nations and among the different regions within our own country.

If we fail to act soon, we will face an economic, social and political crisis that will threaten our free institutions.

But we still have another choice. We can begin to prepare right now. We can decide to act while there is time.

That is the concept of the energy policy we will present on Wednesday. Our national energy plan is based on ten fundamental principles.

The first principle is that we can have an effective and comprehensive energy policy only if the government takes responsibility for it and if the people understand the seriousness of the challenge and are willing to make sacrifices.

The second principle is that healthy economic growth must continue. Only by saving energy can we maintain our standard of living and keep our people at work. An effective conservation program will create hundreds of thousands of new jobs.

The third principle is that we must protect the environment. Our energy problems have the same cause as our environmental problems – wasteful use of resources. Conservation helps us solve both at once.

The fourth principle is that we must reduce our vulnerability to potentially devastating embargoes. We can protect ourselves from uncertain supplies by reducing our demand for oil, making the most of our abundant resources such as coal, and developing a strategic petroleum reserve.

The fifth principle is that we must be fair. Our solutions must ask equal sacrifices from every region, every class of people, every interest group. Industry will have to do its part to conserve, just as the consumers will. The energy producers deserve fair treatment, but we will not let the oil companies profiteer.

The sixth principle, and the cornerstone of our policy, is to reduce the demand through conservation. Our emphasis on conservation is a clear difference between this plan and others which merely encouraged crash production efforts. Conservation is the quickest, cheapest, most practical source of energy. Conservation is the only way we can buy a barrel of oil for a few dollars. It costs about $13 to waste it.

The seventh principle is that prices should generally reflect the true replacement costs of energy. We are only cheating ourselves if we make energy artificially cheap and use more than we can really afford.

The eighth principle is that government policies must be predictable and certain. Both consumers and producers need policies they can count on so they can plan ahead. This is one reason I am working with the Congress to create a new Department of Energy, to replace more than fifty different agencies that now have some control over energy.

The ninth principle is that we must conserve the fuels that are scarcest and make the most of those that are more plentiful. We can't continue to use oil and gas for 75 percent of our consumption when they make up 7 percent of our domestic reserves. We need to shift to plentiful coal while taking care to protect the environment, and to apply stricter safety standards to nuclear energy.

The tenth principle is that we must start now to develop the new, unconventional sources of energy we will rely on in the next century.

These ten principles have guided the development of the policy I would describe to you and the Congress on Wednesday.

Our energy plan will also include a number of specific goals, to measure our progress toward a stable energy system.

These are the goals we set for 1985:

- *Reduce the annual growth rate in our energy demand to less than two percent.*
- *Reduce gasoline consumption by ten percent below its current level.*
- *Cut in half the portion of United States oil which is imported, from a potential level of sixteen million barrels to six million barrels a day.*
- *Establish a strategic petroleum reserve of one billion barrels, more than six months' supply.*
- *Increase our coal production by about two-thirds, to more than one billion tons a year.*
- *Insulate 90 percent of American homes and all new buildings.*

- *Use solar energy in more than two and one-half million houses.*

We will monitor our progress toward these goals year by year. Our plan will call for stricter conservation measures if we fall behind.

I can't tell you that these measures will be easy, nor will they be popular. But I think most of you realize that a policy which does not ask for changes or sacrifices would not be an effective policy.

This plan is essential to protect our jobs, our environment, our standard of living and our future.

Whether this plan truly makes a difference will be decided not here in Washington, but in every town and every factory, in every home and on every highway and every farm.

I believe this can be a positive challenge. There is something especially American in the kinds of changes we have to make. We have been proud through our history of being efficient people.

We have been proud of our leadership in the world. Now we have a chance again to give the world a positive example.

And we have been proud of our vision of the future. We have always wanted to give our children and grandchildren a world richer in possibilities than we've had. They are the ones we must provide for now. They are the ones who will suffer most if we don't act.

I've given you some of the principles of the plan.

I am sure each of you will find something you don't like about the specifics of our proposal. It will demand that we make sacrifices and changes in our lives. To some degree, the sacrifices will be painful, but so is any meaningful sacrifice. It will lead to some higher costs and to some greater inconveniences for everyone.

But the sacrifices will be gradual, realistic and necessary. Above all, they will be fair. No one will gain an unfair advantage through this plan. No one will be asked to bear an unfair burden. We will monitor the accuracy of data from the oil and natural gas companies so that we will know their true production, supplies, reserves and profits.

The citizens who insist on driving large, unnecessarily powerful cars must expect to pay more for that luxury.

We can be sure that all the special interest groups in the country will attack the part of this plan that affects them directly. They will say that sacrifice is fine, as long as other people do it, but that their sacrifice is unreasonable, or unfair, or harmful to the country. If they succeed, then the burden on the ordinary citizen, who is not organized into an interest group, would be crushing.

There should be only one test for this program: whether it will help our country.

Other generation of Americans have faced and mastered great challenges. I have faith that meeting this challenge will make our own lives even richer. If you will join me so that we can work together with patriotism and courage, we will again prove that our great nation can lead the world into an age of peace, independence and freedom.

---

* Jimmy Carter televised speech – April 18, 1977.
Jimmy Carter, "The President's Proposed Energy Policy." 18 April 1977. Vital Speeches of the Day, Vol. XXXXIII, No. 14, May 1, 1977, pp. 418-420.

Courtesy of Jimmy Carter, the American Experience, PBS.

See original at:

http://www.pbs.org/wgbh/amex/carter/filmmore/ps_energy.html

4.
Other Energy Sources

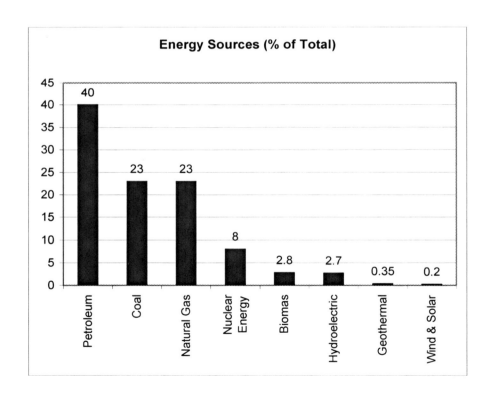

Other Energy Sources

The illustration above tells the whole story. First, it shows the U.S. heavy reliance on the fossil fuels; petroleum, natural gas and coal. Second, the illustration shows the renewable energy share. It shows the insignificant energy share of the two most talked-about renewable energy sources, wind and solar power. Discussion of petroleum replacements to gain energy independence must include discussion of what is presently available and so this section will review all of the alternative sources available today and examine their potential.

Natural Gas

Natural gas is a fossil fuel. It was formed in the same manner as oil. Just as oil, natural gas has been trapped underground beneath layers of sand and rock. Natural gas is a colorless and odorless gas, of which the main ingredient is methane. Natural gas suffers the same fate as oil in that the North American reserves are small compared to the rest of the world, where the Middle East, Europe and Eurasia hold the largest reserves. Unlike oil, most of the natural gas that is consumed in the United States currently comes from the United States. Some of today's supplies also come from Canada. ANWR and Alaska, as a whole, provide for a discussion of potential large natural gas reserves; however, due to political pressures, these reserves along with ANWR's oil reserves remain frozen.

In the United States approximately 23 percent of the total energy comes from natural gas. Natural gas has many uses; it is being used in American homes as heating fuel and cooking fuel. It is being used by natural gas fired power plants to produce a large portion of the U.S. electricity. Production of U.S. electricity uses as much as 23 percent of the natural gas total. It is being used by the U.S. industry, not just for heat but also for many other uses. The industry uses natural gas to produce steel, fertilizers, plastics and glass, along with many other products. It is even used as a fuel to power some of the vehicles driven on the U.S. highways. A chemical called mercaptan is added to natural gas before its distribution, to give it an unpleasant rotten egg-like smell; so the gas can be detected easily.

Of all the fossil fuels, natural gas contains the least amount of carbon and therefore produces the smallest amount of CO_2 emissions. Because of its much smaller carbon content, natural gas is considered a "clean fuel." Coal on the other hand contains the most carbon and is considered the dirtiest fuel.

Natural gas use is expected to grow in the future. Some predictions provide for the usage of natural gas to double by 2025. As the U.S. becomes more dependent on natural gas, it will have to be imported from other parts of the world. The world supplies of natural gas are vast today. The reserves-to-production ratios show some seventy years of supplies available. Eventually though, with increased consumption, the supplies will start to shrink and natural gas will become the same type of trap as the petroleum has become today.

Coal

Coal is another fossil fuel that the U.S. uses for energy. Coal is a brown to black sedimentary rock and is the most abundant fossil fuel found in the United States. In fact, the United States has the largest proven reserves of coal of any country in the world. Approximately 20 percent of the total world's reserves are in the U.S.

Surface mining or underground mining is used to retrieve coal from the ground. Surface mining is used when the coal is buried less than 200 feet deep. Giant machines remove the soil and the rock above it and expose the coal. The exposed coal is then taken to a plant nearby to remove the dirt and rocks; after which it is shipped to the markets. When the mining of coal is complete, the rock and topsoil are replaced and the area is replanted. Surface mining accounts for about two-thirds of the current coal production. Surface mining is easier and cheaper, and therefore more popular than it's counterpart, the underground mines. Underground mining is used when the coal is buried deep in the ground. These mines can be as much as 1,000 feet deep and they rely on conveyor belts to bring the coal to the surface.

The United States production has mostly remained unchanged in the last ten years. Using the current consumption rate, the supplies of coal are estimated to last well over two hundred years. For the most part, coal is used to generate the U.S. electricity and more than 90 percent of coal produced is used for this purpose. It is unlikely that the U.S. is going to be importing coal any time soon and therefore coal is going to be a fuel source that the U.S. will depend on, making its electricity for a long time to come.

It is however a dirty fuel which makes it impractical to be used elsewhere. When coal is burned it gives off carbon dioxide, just as all of the fossil fuels. Of all the fossil fuels, coal is considered to be the dirtiest. The U.S. power plants have been trying to come up with ways to clean the smoke, so the harmful pollutants are reduced in the future.

Nuclear Energy

Martin Klaproth discovered uranium in 1798. In 1938 Otto Hahn and Fritz Strassman discovered that uranium could be split and would release energy in the process. Uranium is abundant to earth. Uranium is very dense; one gallon in volume of uranium weighs about 150 pounds. Uranium boils at 6,900°F. It is found as a high-grade uranium ore, with a concentration of 20,000 ppm U. It is found as a low-grade uranium ore with a concentration of 1,000 ppm U. In the earth's crust it has an average concentration of 2.8 ppm U and in ocean water it has a concentration of .003 ppm U. Currently only the natural uranium ore with the high concentration is considered recoverable. The uranium in the rocks in the Earth's crust and ocean water is not considered recoverable due to the low concentrations. The world's recoverable uranium resources are estimated to be 3.5 million tons of uranium; with the current usage of 68,000 tons per year, a fifty-year supply.

Uranium is much more abundant to the earth than that. According to Professor Bernard Cohen of the University of Pittsburgh, seawater contains five billion tons of uranium, which is enough to supply all of the world's energy for several million years. In addition, rivers constantly replenish the oceans' uranium supply, at a rate sufficient to provide twenty-five times the world's current energy use. And finally, the earth's crust holds enough uranium to satisfy the uranium supply for another five billion years. He claims that this low concentration of uranium is recoverable at $1,000 per pound, where today's cost is $20 to $30 per pound mining the high-grade uranium.

Nuclear energy accounts for about 20 percent of the total electrical power in the United States. The United States has sixty-six nuclear power plants. Nuclear power plants use a process called fission to run their turbines. Fission takes place inside a nuclear reactor. It splits larger atoms into smaller atoms and during this process heat energy is released. Nuclear power plants use fission heat to produce steam, and just as in the coal or natural gas fired power plants, the steam runs the turbines and the turbines run the generators that produce electricity.

Nuclear power plants do not produce carbon dioxide. In fact, the use of nuclear power plants prevents some 700 million tons of carbon dioxide from being released into the atmosphere. The byproduct of nuclear power, however, is nuclear waste. Radioactive waste is the main problem of the nuclear program. It must be stored in special containers underwater for a long time. Plans are being made by the Department of Energy to move this waste to an underground storage facility in Yucca Mountain, Nevada.

No new nuclear power plants have been built in the United States for quite some time now. It is not the radioactive nuclear waste that is the problem; it is the "nuclear, uranium and atomic" words themselves that actually cause problems. Most neighborhoods are generally uneasy to have any type of power plant built nearby but when the word "nuclear" is added to the equation, then building a new plant becomes totally unacceptable if not impossible.

Worldwide, thirty-one countries use nuclear reactors for a portion of their total electric generation. Globally some 440 nuclear reactors supply 16 percent of the world's electricity, with 364,000 MW of total capacity. Even though the U.S. produces one-third of nuclear power, the ratio of total power generation in the U.S. is low, approximately 20 percent when compared to other countries. Lithuania and France's nuclear power generation is approaching 80 percent of the total power generated in those countries. In fact, there are seventeen countries in which the share of nuclear power is higher than in the United States.

The most significant quality of uranium is the amount of energy it contains per pound compared to other sources of energy, specifically the fossil fuels. One ton of uranium can produce 40,000,000 kilowatt-hours of electricity; meaning that one ton of uranium can replace 16,000 tons of coal or 80,000 barrels of oil. Even more amazing is the potential uranium carries after it is used.

Naturally occurring uranium contains .7 percent of the uranium-235 isotope, the energy-producing fissionable isotope, and it contains 99.3 percent of the uranium-238 isotope, the non-fissionable isotope. This means that only 1 to 2 percent of the uranium, depending on how it was enriched, is used to produce this huge amount of energy and 98 to 99 percent of nuclear power remains and is stored as radioactive waste. In a fast breeder nuclear power reactor, a chain reaction converts the non-fissionable uranium-238 to fissionable plutonium PU-239. This fissionable fuel is now able to produce heat. This means that the fast breeder nuclear power reactor can now extract sixty or more times the energy from each pound of uranium than is possible in today's nuclear power reactors.

With this potential in mind, the numbers of 16,000 tons of coal and the 80,000 barrels of oil needed to replace one ton of uranium all of a sudden become insignificant, or just plain meaningless. It is estimated that the stored waste radioactive fuel, if converted to fissionable plutonium fuel, could provide all of the U.S. electric power for the next 1,500 years. Additionally, the vast nuclear weapons stockpiles could provide the U.S. with another "few" years of energy. No more CO_2, no more acid rain and no more pollution.

Breeder reactors are not new. The first experimental breeder reactor was in operation in 1951 in Idaho. This reactor was plutonium fueled and cooled with sodium and potassium. Another fast breeder reactor was built in Lemont, Illinois and started operating in 1963, and a 350 MWe commercial capacity reactor was started at the Clinch River site in Tennessee in the 1970s. Due to the environmentalist pressures, the Carter administration abandoned the Tennessee project and later, in 1992, the Clinton administration shut down the Illinois project. Though the technology exists, no fast breeder reactors operate in the U.S. today.

France, on the other hand, operated the first large-scale fast breeder reactor, named the Super-Phenix, since 1984. The optimum breeding utilizes 75 percent of the energy in the uranium, compared to the 1 percent in the typical nuclear reactors operating throughout the world today.

It has been concluded that the nuclear power plants are safe. Current nuclear power technology provides the most bang for the buck. It is by far the most efficient use of energy today. If fast breeder reactor technology was utilized, the OPEC oil could become obsolete very quickly. However, the thought of an accident, no matter how remote, makes most people uneasy.

It is not unreasonable to expect a little more effort to increase the nuclear share of the overall electric production. To double the current production from 20 percent to 40 percent would require building one hundred nuclear plants. To replace coal altogether, three hundred new nuclear plants would need to be built. This effort would provide for much cleaner air by reducing large CO_2 emissions. However, because of the public fear and the politics that surround nuclear power, it is unlikely that the U.S. is going to see nuclear power expand any time soon. How strange; the only fuel that could potentially provide an unlimited amount of fuel for the U.S. almost effortlessly is being completely ignored.

Wind Energy

Wind is produced when the sun heats up the earth's surface. The sun heats up the various parts of the earth's surface unevenly. The land heats up and cools down much faster than the water surface of the oceans, thus during the day the air over the land heats up more quickly than the air over the water. The warm air over the land rises and the cold air from the ocean moves in to replace it. As the air moves, wind is created. At night this process is reversed. Essentially the same effect is created on a more global scale when the equator receives more of the sun's energy than the northern and southern poles, creating large atmospheric winds circling the earth.

Wind speeds are faster in higher elevations and are the preferred sites for wind farms. Just a small change in elevation can drastically improve the wind turbine efficiency. Seashores on the other hand take advantage of the convection, the difference of heating and cooling of land and the ocean, and provide good sites for wind turbines.

Wind energy uses the energy of the wind to create electricity. Wind turbines are used for this purpose. A wind turbine is a simple design. As the wind spins the propellers, the turning shaft of the propeller turns the generator and the generator produces electricity. Wind turbines depend on wind to produce electricity. The turbines are ready to operate continuously; nevertheless, the lack of wind prevents them to do so. Winds in excess of fifteen miles per hour are required to operate at optimal capacity, and since that is not usually the case, the turbine's capacity usually does not exceed 30 to 35 percent. For that reason, a wind farm needs triple the capacity. A 1,000 MW wind farm would generate only about 300 to 350 MW of electricity.

Presently, wind turbines have a variety of applications around the world. Wind turbines can generate electricity for a single home or a farm of wind turbines can provide power to entire city grids. In remote areas, a village without access to electricity uses wind turbines to power their homes. An electric grid powered by wind meets the need of homeowners who depend on this power to light their homes, charge their batteries and a variety of other applications. The preference is to use smaller turbines since they do not require as much wind and therefore are more dependable to produce power continually.

California leads the way to wind-generated power in the United States. The homeowners in California discovered wind power when the electricity crisis in the 1970s developed. The prolonged crisis has generated increased interest in wind energy generators. Currently California offers rebates to homeowners willing to purchase a home wind energy system.

A small wind generator is installed near a home. The wind system is interconnected with the existing utility power. The homeowner usually uses the electricity which is generated and any excess power is delivered back to the utilities grid. Depending on whether power is needed or a surplus of the power is generated, the utility meter runs forward or backward. This arrangement assures the full use of the wind turbine. Homeowners can typically break even in ten years or less and then have essentially free electricity. About 13,000 windmills produce around 1 percent of California's electric consumption. To put it in perspective, the total production of all windmills amounts to about one-half of a nuclear power plant.

The United States as a whole lags behind the world in wind power generation. Europe accounts for three-quarters of the wind generation production and Germany leads the way. In 2005 wind power accounted for 6 percent of German electric generation and by 2010 Germany plans to expand the electric generation to 12.5

percent. This is contrasted with the United States, where according to American Wind Energy Association only .4 percent of the total electricity was generated by wind in 2005. Denmark, with 20 percent of its total electricity production being wind power, leads the world, illustrating the wind power potential. Both Germany and Denmark are the leading exporters of wind turbines.

New technology and research is going on to drop the cost of energy from wind and to make wind turbines larger and more efficient. A wind farm is far less expensive to build than a conventional electric power plant, and as the demand for electricity increases additional wind turbines can be added. Today, wind-generated electricity is less than 5 cents per kilowatt-hour. With a development of multi-mega-watt wind turbines, the dirty coal-powered electric plants will gain competition. Several electrical providers currently use wind farms to supply a portion of power to their customers.

Wind energy is clean and renewable but it has some drawbacks. With the favorite installation places being in the hillsides and on seashores, the windmill aesthetics spoil the view. Since wind turbines take energy from the wind, they need to be spaced far from each other, and several hundred square miles may be needed to provide the energy of a typical power plant. Therefore they can not be installed near demand centers, thus requiring new longer transmission lines which just add to the cost. The intermittency of wind power is another potential drawback, where a standby power plant has to be kept when wind is not available. And lastly, there are the complaints from the environmentalists who claim that the wind farms kill too many birds.

The overwhelming positive has to be clean, renewable energy. There are no health hazards due to CO_2 emissions, as in the case of coal fired plant, nor is there threat of a nuclear meltdown, creating another Chernobyl. Even though large areas are required for wind farms to collect the energy from wind efficiently, almost all the land remains usable. For example, when a farmer makes his land available for a wind farm he can still plant and farm around the wind turbines. In addition, installation of large 5-MW wind turbines offshore takes no available land space at all. And because a wind turbine is a simple machine, which is more dependent on wind than new technology, one must question why wind energy isn't utilized on a larger scale in the U.S.

Solar Energy

Solar technology uses the sun's energy to provide electricity to heat and light homes, to provide hot water, and many other uses in businesses and industries. The current drawbacks of solar energy are: first, it can only be harnessed when the sun's radiation hits the earth's surface, which means only during a day; second, the area required to collect the energy economically. Solar energy hits the earth's surface at 1.4 kilowatts per square meter. The average power for North America is .125 to .375 kilowatts per square meter or 3 to 9 kilowatt hours per square meter per day.

The best-known technology is the use of photovoltaic solar cell. Photovoltaic solar cell converts sunlight directly into electricity. Use of photovoltaic cells has been limited due to their high costs. Solar calculators and solar watches are familiar low-power devices in use today. The challenge is to make these cells efficient enough to be able to power the entire home and even to power the entire electric grid. Today's photovoltaic cells, which are usually made from silicon alloy, are only 15 percent efficient; therefore the power generated by solar panels is relatively low, 15 to 60 watts per square meter, or .45 to 1.35 KW per square meter per day. Researchers are trying to increase the efficiency to higher levels through the use of different materials and different collecting systems.

A photovoltaic solar cell, or PV cell, is the basic building block. An individual solar cell produces just a miniscule amount of power. PV cells connect together to form modules and modules connect together to create even larger units called arrays. When arrays are interconnected they can provide enough power for any application. They can be used to provide power to a single home, or when they combine together in large numbers they could generate electricity for a power plant.

The major applications today are heating water in swimming pools and water heaters. These systems usually have a flat plate solar collector which is mounted on the roof, facing the sun. The sun heats the plate inside the collector, which is filled with a transporter fluid, usually water or air. Hot water circulates through the tubes, thus heating the water in swimming pools and hot water tanks. Other applications where solar panels may be used is providing outdoor lighting around homes or providing lighting in remote areas such as outdoors for camping or fishing.

Using photovoltaic cells to heat and light homes is still very expensive and although the costs have come down, it still can run $12 to $15 per installed watt. Larger systems can be installed for much less, usually in the range of $3 to $4 per installed watt of power.

Photovoltaic systems convert sunlight directly into electricity without requiring mechanical generators or creating unwanted byproducts. Installed on the roof of a house, the photovoltaic system is hidden away, requiring no additional space. While impractical to replace power plant generated electricity, photovoltaic cells find limited use in remote areas, where no electricity is available.

Japan and Germany lead the way in solar power. Japan's installed capacity is 1,100 MW and it currently uses about half of all the solar modules produced worldwide, primarily for residential applications. Germany's installed capacity is 800 MW. Both countries allow most of the solar-generated power to be connected to an electric grid. This compares with the U.S. installed capacity of 365 MW, with about 50 percent connected to the grid.

Higher intensity light needs to be achieved for commercial applications. The concentrator system collects the sunlight over a large area, sending it to a focal point to create a high-intensity heat source. Point focus uses the parabolic dish to concentrate the light at the point, where a solar cell or some thermal energy receiver receives the light. This type of system uses fewer solar cells, which are the most expensive portion of the entire solar array, thus reducing the overall cost. Line focus uses a "parabolic through," a "curved through," to concentrate the light along a line. Today, concentrated solar power plants use the parabolic through systems to produce electricity. The sunlight is reflected onto a hollow tube above the curved through, heating the fluid within. The SEGS systems in California use oil to carry the heat. The oil passes through a heat exchanger, creating steam, and the steam in turn is used to produce electricity.

To conclude, the photovoltaic solar cell makes it possible to harness the unlimited energy of the sun. One day solar energy will be harnessed on a much larger scale. It is just a matter of time before solar energy replaces natural gas and electric bills in home-heating applications. It is just a matter of time before solar energy will be capable of replacing the gasoline Americans use in their automobiles. It is just a matter of time before the clean solar energy of the sun will be able to replace all of the fossil fuels.

Biomass

Biomass is the leading source of renewable energy in the United States. According to EIA, biomass accounts for 47 percent of the renewable energy or 3 percent of total energy needs. Agriculture and forest biomass byproducts are used to generate electricity. Biomass includes all plant-derived material, which means that biomass is a fully renewable resource, and because it is created by means of photosynthesis, it creates no net carbon dioxide emissions. Burning biomass produces carbon dioxide, which is offset by plants using carbon dioxide to grow, thus no net increase of carbon dioxide entering the atmosphere. There are many types of biomass such as wood, plants and organic industrial wastes. Wood and plant products have been used as an energy source for thousands of years and may find their usefulness still.

Biofuel is a fuel produced from biomass. The two most common biofuels are biodiesel and ethanol. Both are currently derived from agricultural crops and their popularity is on the rise. The primary reason for the dramatic increase in popularity and demand is because biofuels are liquids, which can substitute for liquid diesel and gasoline in cars and trucks. Currently biofuels are only product capable of replacing the liquid fossil fuels.

Biodiesel is a diesel fuel that can be burned in any diesel engine as a fuel or as a fuel additive. Biodiesel can be made from a large variety of oil products. First it is produced from waste vegetable oil and waste animal fat. The second way it is produced is from grown crops such as soybeans, rapeseeds, mustard and palm oil. Unfortunately the current production of biodiesel replaces only about 1 percent of fuel in the U.S.

The most promising production of biodiesel, currently in the experimental stage, is from algae. According to Wikipedia (http://oakhavenpc.org/cultivating_algae.htm), algae can produce 250 times the amount of oil per acre as soybeans (10,000 gallons per acre). Growing soybeans can produce a mere 40 gallons of oil per acre per year. Rapeseed at 110 gallons of oil per acre and even palm oil with 650 gallons of oil per acre per year yield could not even begin to compete with algae production if it ever took off.

Biomass can also be converted into liquid biofuels by the process of fermentation. Fermenting a biomass such as corn, which is high in carbohydrates, produces a liquid alcohol called "ethanol." Ethanol, along with its other names

– alcohol fuel or grain alcohol, refers to a colorless liquid formed by fermentation of sugars. Ethanol can be used in many cars directly, as gasoline or as a gasoline additive. Gasohol, a 10 percent ethanol and 90 percent gasoline mixture, can be used in any current gasoline-powered automobile. Other ethanol mixtures can only be used in specifically converted engines, which are designed to run on different types of fuels. These vehicles are called flexible-fuel vehicles and can run on pure gasoline, pure ethanol, or any mixture of the two. In the United States, flexible-fuel vehicles accept a mixture of 85 percent ethanol, and 15 percent gasoline.

With gasoline above $2 per gallon, the production of ethanol from corn has become a profitable business. One bushel of corn yields 2.7 gallons and one acre of land can yield an average of 140 bushels. That amounts to a little under 400 gallons of ethanol per acre per year.

Ethanol can be made from many different plant matters. In the U.S., ethanol is derived primarily from corn. With advanced bioethanol technology it will be possible to make ethanol from materials such as corn stalks and saw dust. The most promising potential may be in fast-growing grasses such as switch grass (prairie grass), with a potential ethanol yield of 1,000 gallons per acre. With the corn crop yielding less than 400 gallons per acre, one can appreciate the potential of other plants.

Currently the production of ethanol is small. Present U.S. production is growing slowly, from 3.4 billion gallons in 2004 to 4 billion gallons per year in 2005, with a projected 5.5 billion gallons per year in 2006. The current existing production could replace slightly more than 1 percent of the United States total oil consumption. In 2005 the energy policy act required ethanol production to increase to 7.5 billion gallons by 2012, which would be just short of 2 percent of the current total oil use.

Both ethanol and biodiesel reduce carbon monoxide emissions and help to reduce oil dependence. To get serious in providing oil independence to this country, the United States will have to include biofuels as a major part of their overall planning. For now the government only talks of ethanol and biodiesel as being primarily produced from growing plants using up scarcely available farmland, however, once this country gets serious about producing these renewable fuels, then the soybeans with their 40 gallons per acre yield and even the corn with its 400 gallon per acre yield will begin to be replaced by yields derived from algae, which takes up no farmland at all.

Here is another U.S. illogical paradox. Provide farm subsidies to produce 40 gallons of fuel per acre and withhold all funds for a research that could yield 10,000 gallons per acre. One would think that the algae research would have gained the government's priority a long time ago. It almost seems that someone is working overtime not to let this information be known to the U.S. public.

Hydropower

Hydropower captures the kinetic energy of falling water and turns it into electricity. Hydropower is responsible for generating nearly 10 percent of the electricity used in the United States at the present time. It is the largest source of renewable power generation.

Hydropower is called a renewable energy because the water flow is replenished by rainfall and snow.

A hydropower plant stores water from a river in a reservoir. Falling water spins the turbine, which in turn spins the electric generator, thus producing electricity. The amount of electricity that can be generated depends on how much water is available and how far the water drops. The easiest place to build a hydro plant is on a river with a natural waterfall. A natural waterfall supplies the height and when combined with a sufficient water flow volume, provides for excellent electrical generation. The second option for building the hydro plant is to build a dam. A dam creates an artificial waterfall along with a large reservoir of water. Stored water means stored energy. Stored water is released at various rates as electricity is needed. During the day, to accommodate the high demand, water is released at a faster rate; during the night, when the electric demand is low, large gates slow down the flow of water to store the water energy for the next day, when it will be needed again.

Perhaps the two best-known hydroelectric power plants in the United States are the Niagara Falls Power Plant and the Hoover Dam. The potential of Niagara Falls was recognized as early as 1759 when Daniel Joncairs harnessed a small portion of Niagara's energy by building a small canal diverting water from the falls to his sawmill. In 1881, Charles Brush provided electric lights and a generator, and using the electric power supplied by Jacob Schellkopf's water turbines illuminated the streets of Niagara Falls. Brush's generator lit sixteen streetlights using a direct current. Direct current could only be used locally because it cannot be transmitted over long distances. When Nikola Tesla invented an "alternating current power transmission system," which enabled transmission of electricity over long distances, the construction of a power hydroelectric station began. The hydroelectric station was capable of sending 75 MW of power capacity as far away as Buffalo, New York.

Currently the Niagara River provides 2,575 MW of the USA's power with twenty-five generators, and 2,045 MW of Canada's power with thirty-nine generators. The most powerful stations are the Robert Moses Niagara Power Plant on the American side and Sir Adam Beck #1 and #2 on the Canadian side. In 2005, the Ontario Power Generation announced a plan to build another power station capable of generating additional 1,600 MW of power. Combined, the U.S. and Canada form the largest hydroelectric-generated power station in North America.

Because an enormous amount of water is needed to produce this electricity, visitors to Niagara Falls never see the full power of the falls. Approximately 50 percent of the water is diverted from the waterfalls through large channels to the power plants during the day, and as much as two-thirds of the water is diverted at night. These large channels originate as far as 1.6 miles upstream from the Horseshoe Falls. Typical undisturbed flow of water over Horseshoe Falls is 100,000 cubic feet per second, with a peak flow as high as 225,000 cubic feet per second. Between November 1 and April 1 the water flows at the minimum of 50,000 cubic feet per second, day and night. During the season, however, the minimum water flow of 50,000 cubic feet per second is used only at night; and during the day visitors get to enjoy a somewhat higher volume of water.

When a natural waterfall is not available, building a dam creates an artificial one. Hoover Dam is the best-known example. Hoover Dam, a concrete gravity-arch dam which is located in the Black Canyon of the Colorado River on the border between Arizona and Nevada, thirty miles away from Las Vegas, is another well-known hydropower plant. Its construction started on April 20, 1931 and was finished on March 1, 1936. Originally it was named Boulder Dam. It was renamed after Herbert Hoover in 1947, who played a major role in its construction. In 1985 Hoover Dam was designated a National Historic Landmark. Hoover Dam, at 726 feet high, is the second-highest dam in the United States. It generates 2080 MW of power and transmits its electricity to Los Angeles, California, some 260 miles away. Hoover Dam not only provides a large amount of power, it also is a spectacle to its 8 to 10 million visitors each year.

Hydroelectric power plants are expensive to build; however, once they are built they provide for a cheap source of energy since they use no fuel. Because these plants burn no fuel, they are the cleanest source of energy in use. The future of hydro plants however, is unclear. The clean and renewable power supply provided by dams, on one hand, is contrasted with the effect this type of electric production has on the environment. Environmentalists claim that wildlife is being destroyed and therefore the states create a very lengthy procedure to obtain licenses to build new dams.

On the brighter side, dams also provide flood control, swimming, fishing and prime real estate along its edges. Estimates of how many new hydro power plants can be built in the United States vary. One extreme is that all the best places have already been used and there are no more good places to be found. The other extreme is that only 5 percent of the hydropower potential is being utilized; which means that hydropower can provide as much as twenty times the amount of power it provides today. Maybe twenty times is stretching the reality a bit, but just doubling, or tripling the hydroelectric generation from 10 percent to 20 or 30 percent would be a giant step toward a clean, carbon dioxide-free energy source.

Geothermal Energy

Geothermal energy is simply heat from the earth. With depth the earth becomes warmer. Depths of 3 to 5 miles beneath the earth's surface contain a hot rock capable of boiling water. At a depth of 10 miles the temperature exceeds 400^0 F, at depths of 20 miles the temperature reaches 700^0 F, and at the depth of 30 miles below the surface the temperature reaches $1,000^0$ F. This is a temperature which most of the electric power plants use to generate electricity. There is no need to achieve depths of 30 miles; a useful energy to produce electric power can be recovered in depths of 3 to 5 miles, which is technologically achievable now.

The technology to extract this heat involves injecting the cold surface water down to the hot rocks beneath, then to recover the hot water, and finally to extract the heat. Geothermal energy is clean, limitless and renewable. With improved technology it could be possible to go even deeper, taking advantage of much higher temperatures, in which the water recovered could be in the form of superheated steam, heated to hundreds or even thousands of degrees.

Heat recovered in this way could be used as fuel for the electric power plants to produce electricity; in which geothermal energy could sustain the operation of these power plants indefinitely. No need for coal mines, trucks or trains to deliver coal; no need for the use of natural gas and other petroleum products. And if a society determined that nuclear power reactors continue to remain dangerous, the nuclear power could continue in research as geothermal energy takes over. Additionally geothermal energy, just as hydropower, does not produce any pollution.

So why is the geothermal energy not used? Currently, this idea is of no commercial value. Could it be that the U.S. currently lacks the technology? The idea seems much simpler than splitting atoms, as is the case with nuclear power. It also seems to be much simpler than drilling for oil. The oil drillers are searching the planet, on land and in the depths of the oceans. They drill for oil at depths of several miles and then they go through the process to recover it, to refine it, and finally sell it at the end; yet many times they are still unsuccessful at finding oil, thus wasting millions of dollars. This difficult journey compares with geothermal energy as follows: Select a site anywhere in the country close to water, a river or a lake; drill a series of wells with a certain assured success of "striking heat," and recover the heat energy virtually forever. If the same effort was put to drilling for geothermal energy as is put forth to drill for oil, there would be no need for oil. For now however, for whatever reason, this energy source remains futuristic.

Energy Use Projections

The projected production of various forms of energy in the U.S. is expected to grow significantly. First, according to EIA estimates, domestic production of petroleum in the U.S. is expected to continue to decline while the increased demand for petroleum will continue. Secondly, EIA projects that the dirty coal will pick up the slack. The production of coal is expected to nearly double from 20 to 35 quadrillion Btu. Thirdly, the production of natural gas, nuclear and hydropower is estimated to remain essentially unchanged. Finally, there is very little expectation for the use of renewables. Even though the expected production of renewable energy is expected to increase by some 50 percent, it will increase from 5 to 7 quadrillion Btu and so it will continue to dwarf the production and use of fossil fuels. The actual ratio of energy produced by renewable sources compared to fossil fuels will continue to remain the same, in the single digits.

To summarize what has been uncovered here is to conclude that the doomsayers are wrong. The future energy sources are plentiful. Humanity has used wood for thousands of years as the only energy source. For the last two hundred years fossil fuels replaced the use of wood and the future is about to embark on new sources of energy. There are certainly plenty of new exciting sources to choose from; nuclear, hydropower, solar, wind, geothermal and even today's biomass could emerge as a new leader. They could emerge as an energy source that the world could depend on for many thousands of years.

What has been discovered in this section is that the use of petroleum could soon become as strange as

Other Energy Sources

burning wood and coal to heat the homes of today. The question of "how could humankind have ever existed without oil," will be replaced with "how could humankind have ever existed with oil?" The rest of the chapters will deal with how these sources of energy can replace the petroleum products. While the path seems straightforward it will not be without sacrifices. During a transition, Americans may need to pay a little more at times, but at the end of the process, what emerges is a forever source of energy that will cost pennies on the dollar compared to what Americans pay today.

For more information on renewables go to:

http://www.eia.doe.gov/
http://en.wikipedia.org/wiki/Coal
http://en.wikipedia.org/wiki/Nuclear_power
http://en.wikipedia.org/wiki/Natural_gas
http://en.wikipedia.org/wiki/Wind_power
http://en.wikipedia.org/wiki/Solar_cell
http://en.wikipedia.org/wiki/Biodiesel
http://en.wikipedia.org/wiki/Ethanol
http://www1.eere.energy.gov/biomass/ethanol.html
http://www.cobweb.net/~bug2/rock7.htm
http://www.truthaboutenergy.ocm/Fast%20Breeder.htm
http://www.world-nuclear.org/
http://www.nuc.berkeley.edu/thyd/ne161/shir/project5.html
http://www.nrel.gov/gis/solar.html
http://en.wikipedia.org/wiki/Biofuel
http://en.wikipedia.org/wiki/Algaculture

5.
Hydrogen Solution (The Iceland Way)

Hydrogen vs. Electric

*I*n addition to the U.S. natural resources and renewable resources there is also electric and hydrogen. Unlike the natural resources, neither hydrogen nor electric power occurs in nature on its own. Both hydrogen and electric are energy carriers, which means that other sources of energy, specifically the fossil fuels, have to be burned to produce them. This section will discuss which one is likely to emerge on the forefront, to lead the way to the fossil fuel replacement, or if both can coexist side by side.

Both have excellent potential. Hydrogen could virtually replace all of the fuels in existence today. It can

Hydrogen Solution (The Iceland Way)

exist in gaseous form and can be burned similarly to natural gas; it can exist as a liquid, to be burned as gasoline and through the use of fuel cells it can produce electricity. Electric can be used to do all of the above, just a little bit differently. It can be used in electric furnaces and electric stoves to replace natural gas and it can be stored in batteries and used in electric-powered automobiles, thus replacing gasoline.

Both of these energy carriers, however, have flaws. The major flaw of hydrogen is that in gaseous form it needs a large amount of space to be stored, and as liquid it needs to be stored cryogenically in extremely cold temperatures (-252^0C or -423^0F). Additionally, the liquefaction process requires a large amount of energy to cool the hydrogen down to be used as a liquid, and it requires energy to keep this hydrogen in this super-frozen environment. Still, even in liquid form, approximately four times the volume is required to deliver the same amount of energy as gasoline, for example. Storage problems are also the major flaw of electricity. An automobile requires 20 to 30 car batteries to drive 50 to 80 miles, at which time the batteries need to be recharged. This is a cumbersome and time-consuming task to say the least.

Which technology is likely to lead the way? Both technologies are being discussed equally. If both technologies were at an equal developmental stage, then it may be a toss up; the hydrogen economy may even have an edge in that the fuel cells are much lighter; the hydrogen fuel itself weighs hardly anything. This makes the overall weight of a vehicle lighter, when compared to an electric-powered vehicle using battery packs weighing in excess of 1,000 pounds. Technologies, however, are not similarly advanced.

Hydrogen is an unknown choice and is regarded by many as dangerous. The storage problems are still in an experimental stage, and if hydrogen is going to be used as intended, then the fuel cell, which uses hydrogen as a fuel source, needs improvements too. Not just technologically but economically. Electrical energy, on the other hand, is well-known by everyone and it is regarded as safe, and for all practical purposes, it is ready now.

There is a good chance that electric energy will take over the role of replacing the fossil fuels completely on its own, especially if a better electric storage media is found, never giving hydrogen a chance. On the other hand, once the hydrogen and fuel cell kinks are resolved, hydrogen could be the fuel of the future; replacing not only the fossil fuels used today in homes and automobiles, but also replacing the fossil fuels electric power plants use. The most probable scenario may be that even though the electric will begin the process, society will eventually find application for both. Nuclear, geothermal and all the renewables will be used to crank out electricity around the clock, providing electric power to American consumers to use in all of the familiar applications; and then during the off-peak hours the excess electricity will be used to produce hydrogen for use in applications specifically designed for hydrogen use. Let's more closely examine the two sources separately: hydrogen vs. electric.

Hydrogen Fuel

Hydrogen is the simplest, lightest and one of the most common elements in the universe, and it is regarded as the most promising alternative to fossil fuels. On earth, hydrogen does not occur naturally and is almost always combined chemically with other elements such as oxygen and carbon. The most abundant supply of hydrogen is found in water, where it is combined with oxygen. Hydrogen is also found in other organic compounds, mainly hydrocarbons, which include natural gas, oil, methanol, coal and gasoline.

Hydrogen is abundant throughout the world without any borders, and not just a few selected places like oil is. Once extensive research and development finds new innovative ways to produce hydrogen efficiently, it can be produced anywhere, thus eliminating the fossil fuel use disparities that exist today throughout the world. No longer would just a few control the fate of the world's energy. Before too much excitement sets in, let's examine some of the other hydrogen facts.

Most of the hydrogen produced at present is from fossil fuels, natural gas, oil and coal. Currently nearly half of hydrogen is made from steam reforming natural gas. At temperatures of 700°C to 1,100°C in the presence of a catalyst, steam reacts with methane and other hydrocarbons present in natural gas and produces hydrogen. Carbon monoxide is produced as a byproduct. Steam reforming of natural gas is the most common and most cost-efficient production method of hydrogen. The heat required for steam reforming production is made by burning more of the natural gas.

When a very pure form of hydrogen is required for industrial purposes, it is produced from water by a process of electrolysis. Only about 4 percent of hydrogen is made this way. Electrolysis uses electrical current to split water to produce hydrogen. Hydrogen released from water forms a new bond and stores the energy from the electrical current. Electrolysis requires large amounts of electricity to produce hydrogen. Extensive research is going on today to find a cost-effective way to separate hydrogen from water. Once technology enables production of hydrogen using electrolysis by means of renewable electric sources, such as solar and wind, it should be possible to produce unlimited amounts of environmentally friendly hydrogen that could one day supply clean energy to the whole world.

Production of hydrogen is growing rapidly. The U.S. currently produces 11 Mt per year, with a thermal energy of 48 GWt. Worldwide production is 50 Mt per year and the demand is projected to double by 2025. To meet the future demands, thermochemical electrolysis of water using nuclear power reactors is gaining commercial interest. This process uses heat rather than electricity to split water into its components, and since the input is heat rather than electric energy, the claim is that it can be done much more efficiently.

According to UIC Nuclear Issues Briefing, the evolution of hydrogen production using nuclear power is: first, to use off-peak capacity of existing nuclear power plants to drive electrolysis to produce hydrogen, and secondly, within three decades or so, to use nuclear reactors exclusively dedicated to the production of hydrogen, in which nuclear heat will be used to split water directly in high-temperature thermochemical production. These thermochemical processes require very high temperatures (800 to 1000ºC) to achieve high production efficiencies. At these temperatures, nuclear reactors are

Hydrogen Solution (The Iceland Way)

capable of achieving as much as 50 percent efficiency, compared to the current efficiency capability of just about 25 percent.

Once the hydrogen is made it can be burned as a natural gas in cooking stoves. It can also be burned in furnaces to heat homes and businesses. In a liquid form it can be used and burned as gasoline to power automobiles. It can be used as a fuel for generators to produce electricity for all of the electrical applications such as lighting homes and running all of the electric appliances. It can be used commercially in specially modified gasoline and diesel engines to power trucks, planes, ships and even rockets. Finally, it can be used as a fuel in a "fuel cell" producing electricity, creating water as the only byproduct. In all, hydrogen has the potential to replace all of the fossil fuels of today.

NASA has used hydrogen as a rocket fuel since the 1940s. In space, onboard the shuttle, they use hydrogen as a primary fuel. Through the use of a fuel cell combining hydrogen with oxygen, the shuttle produces its onboard electric power while producing drinking water at the same time.

The hydrogen molecule has two hydrogen atoms, thus the H_2. Hydrogen is odorless, colorless and tasteless. Pound for pound, hydrogen contains about three times the energy of gasoline, the highest energy content of any known fuel. Hydrogen's energy value is 60,000 Btu per pound compared to gasoline's energy value of 18,900 Btu per pound. One gallon of gasoline, which weighs 6 pounds, can be replaced by 2.2 pounds of hydrogen. A cubic foot of hydrogen weighs .00264 pounds under standard conditions and a commercial steel hydrogen cylinder pressurized to 2200 pounds per square inch holds approximately 1 pound of compressed hydrogen, thus two large steel cylinders filled with compressed hydrogen gas are needed to replace one gallon of gasoline in automobile gas tanks.

This fact recognizes the biggest problem with hydrogen; it occupies a large amount of space. Despite compressing the hydrogen in highly pressurized cylinders, the volume is still much larger when compared to gasoline, creating a problem with onboard storage. A variety of storage devices are being experimented with. One of the ideas is to increase the pressure in the hydrogen containers, which requires compressors that need fuel to operate, therefore reducing the overall end product efficiency of hydrogen.

Another proposed storage solution is to use ammonia (NH_3) as a hydrogen storage device. This chemically stored hydrogen can then be released through the use of a catalytic reformer. Metal hydrides such as boron hydrate or lithium hydride offer another type of storage. This type of storage creates a volume difference on the order of three times larger when compared to standard gasoline; however, when compared to bulky pressurized steel containers they may offer an attractive alternative.

Finally a problem emerges of how to get the hydrogen to the customer. A solution has been offered to solve this dilemma. It was to provide gasoline reformers to existing gas stations to reform the gasoline in place, thus using the existing infrastructure, and selling hydrogen similarly to the way gasoline is sold today. The hydrogen atom, being the smallest element on earth, can escape through valves having the smallest leaks, and therefore the absolute necessity of having tight seals creates yet another difficult problem which researchers need to address. Because hydrogen is a flammable gas, the primary concern is safety, and until the American population is assured that hydrogen can be handled safely without any worry of ignition, hydrogen is unlikely to gain acceptance.

Once all the problems of hydrogen handling are overcome, then the most promising, and currently the most talked-about use of hydrogen fuel will be powering the "hydrogen fuel cell."

Fuel Cell

A fuel cell is a device that uses hydrogen as a fuel, in which hydrogen combines with oxygen and produces electricity. Fuel cell consists of an anode, a cathode and an electrolyte membrane. Pure hydrogen is supplied to the anode side of a fuel cell, where it comes in contact with a platinum catalyst, which creates a chemical reaction creating an unstable form of hydrogen, positively charged ions (protons), and negatively charged electrons. Positively charged hydrogen protons pass through the electrolyte membrane barrier to the cathode side. The negatively charged electrons travel to the cathode through an external circuit producing electric current. At the cathode, both the hydrogen protons and the electrons combine with oxygen to form water. As long as the hydrogen is supplied to one side of a fuel cell and air is provided to the other side of the cell, then electricity will be produced. Fuel cells works very much like batteries, but unlike batteries they never need recharging.

Fuel cells have many problems today. First, they are too large and bulky for automobile applications. Second, the costs associated with the fuel cell compared to the gasoline engine are large, as much as ten times more, and because fuel cells are estimated to last around 30,000 miles maximum, these costs may be multiplied when compared to a gasoline engine's life expectancy of over 100,000 miles. Third, fuel cell technology is new and untested. Americans have become dependent on gasoline engine technology, which has been perfected to run with minimal maintenance. Today's fuel cell technology provides for a product with many potential problems to be resolved. Fourth, the material component of the fuel cell, the supply of platinum, which is a necessary component of a fuel cell, is in a limited supply and is extremely expensive. With demand for platinum increasing, the price of platinum may make the cost of fuel cells increase even further.

American automakers have been experimenting with fuel cell automobiles for more than a decade now, but they have yet to come up with a design to compete with the gasoline automobile. Currently, hydrogen car prototypes use tanks of hydrogen compressed to 5,000 pounds per square inch (psi), having enough fuel for a maximum range of around 200 miles before they have to refuel. Manufacturers are designing tanks capable of even higher pressures, up to 10,000 psi in some cases, storing more hydrogen fuel for longer trips. GM is currently toying with designs using these pressures. With these kinds of pressures, safety must be of the utmost concern. To get a feel for the pressures these containers have to withstand, consider a typical automobile tire being pressurized to between 20 and 30 psi. Another idea, offered by Chrysler, is to provide a gasoline reformer onboard the car itself, in which the fuel cell vehicle will pump hydrogen-rich fuel such as methanol, natural gas or gasoline, and process the fuel in the onboard reformer, extracting the hydrogen. Hydrogen can then be fed to the fuel cell, powering the automobile.

Hydrogen Economy (The Iceland Way)

Iceland has announced its plan to become a hydrogen economy by 2050. Iceland is rich in renewable natural resources. Hydropower and geothermal power are abundant to this country. Thus, Iceland is able to commit these resources, essentially unlimited natural resources to that country, to the production of hydrogen. How can the U.S. compare?

The U.S. government has recently started to promote hydrogen fuel cells. On January 28, 2003, President Bush announced Hydrogen Fuel Initiative, in which he proposed to transform the U.S. automotive fleet from gasoline to hydrogen fuel. The government claim is that when combined with the Freedom Car initiative, which is aimed at developing affordable hydrogen-powered fuel cell vehicles for mass production, quoting President George Bush, "the first car driven by a child born today could be powered by hydrogen and pollution-free." This extremely optimistic scenario is being openly and extensively questioned as creating suspicion of an alternative motive. As exciting as these forecasts are, they just strengthen the doubts of the nonbelievers.

These concerns are:

1) If hydrogen is proposed as a fossil fuel replacement, why then is it being produced from the same fossil fuels it is supposed to replace? (Specifically natural gas and gasoline, both of which are already in short supply.)

2) The production of hydrogen by electrolysis of water uses large quantities of electric power. This electric power, which is being produced by the very inefficient process of burning hydrocarbons, is then used to produce hydrogen, again considerably reducing the efficiency at the end of the line. Finally, the hydrogen has to be turned back into electricity through the use of a fuel cell, reducing the overall efficiency even further. Comparing apples to apples, the energy transmitted by hydrogen is only half that of electricity at the end use. Also, a lot of fossil fuels have to be burned, releasing a lot more CO_2 to produce hydrogen.

3) If most estimates suggest that twenty to thirty years are needed to complete research and development and another twenty to thirty years to transform to the hydrogen economy, then what is the U.S. government's true objective, specifically when they know that the petroleum replacements will be needed much sooner?

The U.S. government is proposing an extremely optimistic scenario of twenty years or less. Many suspect that these optimistic views provide for not needing to seriously increase the CAFÉ standards of the existing automotive fleet, thus gratifying the oil and automotive industries. Nonbelievers concede that the whole idea of hydrogen production and a hydrogen economy would have a lot more merit if hydrogen was produced using electrolysis of water and if the source of electric power was renewable energy, specifically wind and solar power.

However, with the research needed in the wind and solar power areas, including the extensive research needed in the hydrogen production and storage, the infrastructure for hydrogen distribution and developing the efficiency and reliability of the "fuel cell" itself, the belief is that commercial hydrogen economy is not likely in the near future; certainly not within the government's confident optimistic estimates.

There's no doubt that fuel cell application could someday produce a "hydrogen economy," in which everything that runs on fossil fuels today runs on hydrogen-powered fuel cells. Imagine this scenario of the U.S. transportation fleet as an example: You fill up your automobile with water. Sunlight radiation through the use of highly efficient solar cells extracts hydrogen from the water through the process of electrolysis. Hydrogen is then fed to the fuel cell to generate electricity to power your vehicle. No other fuel is necessary. And because the electrolysis of water was done using the sunlight radiation itself, no harmful greenhouse gases were released to the atmosphere. At the end of the whole process the car tailpipe releases clean water.

Hydrogen may eventually be produced in unlimited quantities, where the rest of the world could be brought onboard to have standards of living similar to the standard of living in the U.S., thus creating markets rich enough to buy American-made products. For now fuel cells face technological and economical obstacles, but the eventual potential is enormous. Maybe one hundred years from now hydrogen will have its chance and the hydrogen economy will come to pass.

The intent here, however, is to come up with a potential substitute for fossil fuel, particularly the transportation fuel, to become energy-independent in one generation, in twenty years or less. Hydrogen as an energy carrier has a long way to go to become a reality. At least a generation is needed to overcome technological obstacles to produce hydrogen efficiently. There is also a familiarity and safety hindrance, in which significant time (possibly a generation or more) will be needed for the population to become familiar with the hydrogen concept and to accept it as a safe and familiar everyday fuel. And finally, the implementation obstacle, the transformation process from gasoline to hydrogen itself, is likely to take another generation.

The overall process will be lengthy and difficult, and no doubt the time it takes to produce a hydrogen economy will take longer than one generation. Therefore, unless some significant revolutionary breakthrough happens to speed up hydrogen technology, in which the hydrogen transformation can be implemented in a shorter time than is currently being forecasted, it will not be discussed as a near-term potential replacement here.

For more detailed information go to:

http://www.fueleconomy.gov/feg/fcv_PEM.shtml
http://en.wikipedia.org/wiki/Hydrogen
http://en.wikipedia.org/wiki/Hydrogen_economy
http://en.wikipedia.org/wiki/Fuel_cell
http://en.wikipedia.org/wiki/Hydrogen_vehicle
http://www.uic.com.au/nip73.htm

6.
Electric Solution
(The Brazil Way)

Introduction to Electricity

*E*lectric power, like hydrogen, is a secondary energy carrier, which means that another form of energy such as coal or natural gas has to be burned to make it. With the invention of the light bulb by Thomas Edison and the alternating current pioneered by Nikola Tesla, electricity became the most widely used form of energy in the U.S. Today's civilization takes electricity for granted. Electric power is taking over a bigger and bigger share of the total energy use.

Most of the appliances in homes run on electricity. TVs, VCRs and DVD players all use electricity. All of the new gadgets such as computers, printers, scanners and cell phones that are used at home and in the business world run on electricity. The computer-era, Internet-era, and laser-era would not exist without electricity. Finally, any new device, tool or appliance or any other imaginable contraption invented in the future is likely going to use electricity as a power source. The demand for electricity is continually increasing and it is predicted by many that electricity may one day power everything.

This widespread use of electricity has positioned electric power far ahead of hydrogen as a potential candidate to replace oil. Electricity is currently being discussed as a possible candidate to replace the gasoline in automobiles, in which the combustion engine will be replaced with an electric motor using electricity as the power source. Compared to hydrogen, the electric industry and the corporations have demonstrated the ability to produce electric power efficiently and usually in sufficient quantities to meet even the highest demand. Americans do not have to discover what electricity is or if it is a safe product, since everyone has come into contact with it somewhere. Unlike with hydrogen, the lengthy time needed for electric power research and development is over. What remains is the transformation phase itself. How can oil be replaced with the electric power that is used today? Can enough electric capacity be generated to accomplish this task?

It is not the intention to replace all oil with electric power. The idea is, rather, can electric power help the U.S. become energy, specifically oil, independent? Can electric power provide enough energy to replace the oil imports? Could Americans ever accept the transformation to an electric automobile? Can it be done efficiently? And most important of all, could it be done in such a way so it does not significantly increase the transportation budgets of American families or significantly change their lifestyles? This section will concentrate on the light transportation fleet; the automobiles, pickups and SUVs, which use nearly two-thirds of the total oil imports. Before proceeding further, a few electric terms should be defined.

Electric power is the amount of work done by an electric current in a unit time, or a simple rate at which electric energy is consumed or supplied. Power is measured in units called watts (W). Different devices consume different amounts of power. A flashlight may use 1 watt of power while a 100-watt light bulb needs 100 watts of power to operate properly. Space heaters may use 500 to 1,000 watts of power, turning electric power into heat.

A kilowatt (kW) is 1,000 watts.

A kilowatt-hour is the amount of energy used up by a one-kilowatt device in one hour, thus a 100-watt light bulb will use one kilowatt-hour of power in ten hours. Power plants typically use megawatts (MW) to express their output capacities.

Megawatt (MW) is 1,000 kilowatts.

Electric Availability

A major part of the planning process to eliminate oil dependence is to provide an alternative fuel. This source has to be dependable, widely available, commonly accessible, and in abundant quantities to prevent sharp price fluctuations and sharp price increases in the future. Currently, electric power is the only fuel that meets all of these requirements. All of the fossil fuels, coal, natural gas and petroleum can be utilized to produce electric power. In addition to fossil fuels, nuclear power and all renewable sources of energy focus on electric production. With such a wide variety of fuels available to produce electric power, the U.S. industry could find it fairly easy to increase the electric power supply as needed. The industry can start adding additional power plants, or in the case of renewable energies, adding solar farms, windmill farms or any one of the rest of the sources, as the need arises.

This relatively easy obtainable power supply provides a way for the transformation from gasoline-powered combustion engines to electric-powered motors, the easiest and simplest program out there. Coal along with nuclear power will likely continue to provide the majority of the electric power needs. Hydropower, geothermal and the renewable energy sources are good candidates to contribute significantly in the future. Solar, wind and biomass are potential candidates to become leaders in the overall electric production expansion in the near future. On the whole, the expansion of electric power to fuel automobiles is well within the ability of the U.S. industry.

To build a new large power plant requires comparatively short time frames of two to three years for a natural gas plant, three to five years for a coal-fired power plant and five to ten years for a nuclear power plant. Therefore any of these options can be easily adopted to providing additional power as needed. A new modern coal power plant can cost $1,300 per kW or $1.3 billion for a typical 1,000-megawatt plant. Hydroelectric and nuclear power plants are priced similarly. Wind farms and solar farms take much less time to build. Because solar panels and wind turbines can be added one at a time as the increased capacity is needed, the costs associated with adding more of them are, in reality, minimal.

Fossil Fuel Power Plant

A fossil fuel power plant converts the energy stored in fossil fuels. It converts it first into thermal energy, then mechanical energy, and finally into electric energy which is distributed to customers. Fossil fuels such as coal, natural gas or oil are used to boil the water and produce superheated steam under very high pressure. This highly pressurized superheated steam ($1,000°$ F) is fed into a turbine; it hits the turbine blades and spins the turbine. The shaft of the spinning turbine turns the generator and a system of magnets, where electricity is generated, thus the mechanical energy is converted into electrical energy.

Transformers, developed by George Westinghouse, are used to move electricity to where it is needed. A transformer provides the means so the electricity can be transported efficiently over long distances. As the produced electricity leaves the power plant it travels along cables to a transformer. The transformer changes the voltage from low to high, allowing the electricity to be transferred over long distances more efficiently. Transition lines transport electricity to substations, where it is changed back to low voltage. Distribution lines transport the electric power to the homes and businesses.

The U.S. produces 3.9 trillion kilowatt-hours of electricity per year. An astounding 49.8 percent of electric power is produced using coal, 19.9 percent using nuclear power, 17.9 percent using natural gas, 6.5 percent using hydroelectric and 3 percent using petroleum. Finally, 2.3 percent of electricity is produced by other renewable sources. Other renewable sources are geothermal power, solar power, wind power and biomass. This provides very realistic yet sad statistics and shows that if solar power, wind power and ethanol are to provide a significant portion of electric power in the near future, then there is a lot of work to be done. The individual fuel potential contributions will be discussed later in a section entitled "Fun with Numbers."

Problems with Electricity

Of course electric power is not without its own problems. The first problem is that most power plants are very inefficient in electricity production. In fact, their efficiency is 35 percent at best. This means that in order to generate 1 Btu of electric power, 3 Btu of energy derived from fossil fuels is used. This makes electric power the most expensive fuel in use currently, but the convenient availability of electricity makes it all worthwhile. Consider these statistics: of the total 100 quadrillion Btu of energy used in one year in the United States, 38 quadrillion Btu is used on electricity production. Of that 38 quadrillion Btu, 12 quads is actually delivered to customers and used and 26 quads is lost in the electric production. Considering that the U.S. uses 23 quads of coal per year,

23 quads of natural gas per year, 40 quads of oil per year and 8 quads of nuclear energy per year, that is some waste.

The second problem with electric is the electric distribution grid, the infrastructure needed for electric distribution. The electric grid in use today has been in place more than fifty years in many places and it has not been designed to distribute today's ever-increasing power demand. Since deregulation of electricity in the 1980s, essentially no new money has been invested in upgrading the electric grid. Whenever a grid failure occurs, politicians regularly revisit the subject and promise money for new improvements. It always seems like once the emergency subsides, the electric grid quickly loses its priority. Therefore the reliability of electric power is not just a function of whether it can provide enough electric power to the customers, but if it can provide power to the customers without any major disruptions. Upgrading the entire electric grid is a major undertaking; it will be expensive and will take many years to complete. Because the electric demand is increasing at a fast pace even without the demand that the auto industry transformation would require, the problem needs to be addressed without any additional delays.

The third problem, the largest problem haunting the electric industry from taking over, is electricity storage. One U.S. gallon of gasoline stores 115,000 Btu or 33,700 watt-hours of energy. A standard 12-volt lead-acid car battery rated at 100 amp-hours delivers 1,200 watt-hours of electric energy and then needs to recharge. With this kind of statistic one can easily see the problem electric power storage is currently facing.

None of these problems are problems that need a lengthy time for research and development. Americans have learned to live with the low-efficiency production of electric power, in which power plants are capable of producing sufficient quantities of electricity to meet the demand. The electric distribution grid does not have to be redesigned requiring new technologies; it just needs to be upgraded. It requires the will of the U.S. politicians and the electric companies to make a decision and invest a few dollars. While the question of large-scale storage of electric power creates a problem needing additional research and development, it in no way means that the transformation program has to be postponed and can not commence right now.

To conclude, even with all its problems, electric power is ready to go. Since battery storage may provide some hindrance during the beginning stages, being the weak link in the electric design, let's examine the battery in greater detail.

Batteries

Battery is not a new concept. They have been used to start automobiles since the early stages of the automobile revolution at the beginning of the 20th century. Portable battery power is used in many applications and is growing in popularity, as new portable applications are being introduced. Portable radios and CD players, watches, laptops and of course the "cell phone," are among the common portable gadgets that currently use batteries. Life without them is hard to fathom. A lot of attention has

Electric Solution (The Brazil Way)

been given to new designs to make the batteries smaller and more portable. New combinations of new materials are being experimented with to increase the battery capacity and make them more durable.

While many improvements to the battery have been introduced for many different individual applications, they have not been considered for use as a large-scale storage device in which a lot of power needs to be stored to provide a substantial amount of energy later; such as with an electric storage media to provide backup power to homes and businesses in case of blackouts or to assist utility companies in providing additional power during the peak demand. Nor have they been considered in harnessing solar and wind power for distribution later, thus eliminating the undesirable effect of intermittency.

The automobile battery has been perfected to start an automobile at any weather and to last the automobile's lifetime, but no consideration has been given so far to battery designs that could be used to power electric automobiles. For now, the electric automobile industry relies only on existing technology and existing designs. They have to experiment by combining as many as fifty batteries to use in the new modern-design electric automobiles.

Batteries have three parts: an anode, a cathode and an electrolyte. The anode and cathode are the negative and positive ends of a typical battery. In a car battery the two heavy lead posts act as terminals. The chemical reaction inside a battery causes a buildup of electrons at the anode and creates an electrical difference between the anode and the cathode. The electrolyte prevents the electrons from going straight from the anode to the cathode but when a wire is connected between the negative and positive terminals, the electrons will flow from one side to the other side as fast as they can to complete the electron deficient reaction. This path is called a circuit and it normally has some type of load connected. The load might be a light bulb, motor or a radio.

There are many types of batteries. There is a zinc-carbon battery with zinc and carbon electrodes and an acidic paste as the electrolyte. Another type is the alkaline battery, with zinc and manganese-oxide electrodes and an alkaline electrolyte. There are also the lithium photo battery, the nickel-cadmium battery, the lithium-ion battery and many others. All of these provide the same function. Only the anode, cathode and the electrolyte, into which the anode and cathode is submerged, are made of different materials.

<u>Lead-acid batteries</u> are the most commonly used rechargeable batteries today. A 12-volt car battery combines six individual cells and the characteristic voltage is 2.1 volts per cell. The cell of a lead-acid battery has one plate made of lead and the other plate made of lead dioxide. Both are immersed in a strong sulfuric acid solution. When the battery is being discharged, the lead combines with the sulfuric acid to create lead sulfate, hydrogen ions in a solution, plus excess electrons. Lead dioxide reacts with the ionized sulfuric acid and the available hydrogen ions plus the electrons from the lead plate, creating lead sulfate and water. As the battery discharges, both plates build up $PbSO_4$, (lead sulfate) and water builds up in the sulfuric acid solution. To simplify this, both plates turn into lead sulfate and the electrolyte turns into water. Eventually these electrochemical

Electric Solution (The Brazil Way)   155

processes change the chemicals in an anode and a cathode to stop supplying the electrons and the battery is discharged.

This reaction in the rechargeable lead-acid battery is completely reversible. When the battery is being recharged, a current is applied to the battery and the flow of electrons is forced backward. The lead sulfate combines with the water to re-form the lead and lead dioxide on the electrodes, restoring the anode and cathode to their original state so the battery can be reused again, providing its full power. The lead-acid batteries are relatively low-cost and can supply high-surge currents when needed in starter motors.

Battery's internal resistance determines how many amps the battery can reliably provide. Batteries are usually built for specific purposes and they differ in construction accordingly. There are two types of applications that manufacturers build their batteries for, starter batteries and deep cycle batteries. Starter batteries are meant to get the combustion engines going; they have many thin lead plates which allow them to discharge a lot of energy in a short amount of time. These batteries do not tolerate being discharged deeply; they are designed to discharge no more than 20 percent of their capacity. Deep cycle batteries differ in that they have thick lead plates which tolerate deep discharges but cannot dispense the charge as quickly.

The capacity of a battery is measured in amp-hours. This is the number of amps the battery can reliably deliver at a reasonable discharge rate for that battery, and for how many hours it is expected to deliver those amps. Most batteries are rated in an electrical capacity for a discharge rate of 20 hours. A 20 amp-hour battery should provide one amp of current for 20 hours before being fully discharged. A standard small car battery has a capacity of 45 amp-hours, meaning it will provide a little over two amps for 20 hours.

If short-circuited, a battery can deliver up to 300 amps. This is what happens when a car is being started. The battery survives because the large cranking loads are short lived. It is quite reasonable to discharge the battery at a much faster rate than its amp-hour rating. However, the demand of a higher current will cause extra losses and the amount of power supplied before the battery is discharged will be less.

AMPS x VOLTS = WATTS

Take, for the sake of example, a standard 12-volt car battery rated at 100 amp-hours. This means it can deliver 5.00 amps for 20 hours. 5.00 amps x 12 volts = 60 watts for 20 hours, or 1,200 watt-hours.

<u>Batteries wired in series</u>

When two 6V, 100Ah batteries are wired in series, the voltage is doubled but the amp-hour capacity remains 100Ah. Total power is 100Ah x 12V = 1,200 watt-hours.

<u>Batteries wired in parallel</u>

Two 6V, 100Ah batteries wired in parallel will have a total storage capacity of 200Ah at 6V. The total power is 200Ah x 6V = 1,200 watt-hours.

Battery Electric Vehicle (BEV)

Electric vehicles were common during the early 1900s and were sold in similar quantities as gasoline engine automobiles. With the improvements to the internal combustion engine, the electric vehicles became undesirable. The primary success of the gasoline-powered vehicle was the cheap and abundant gasoline supply. With the large new oil discoveries at the start of the 20th century, electric vehicles lost their attractiveness and by the 1920s became all but obsolete. Never again have electric vehicles emerged as a serious competitor to the internal combustion engine.

The battery electric vehicle, as the name implies, uses electric power stored in rechargeable batteries as the main source of fuel, and when the fuel runs out the batteries need to be recharged. The disadvantage of a battery electric vehicle is its limited range. Lead-acid battery vehicles are capable of around 80 miles maximum range before they have to be recharged. The more realistic figure is around 50 miles. Electric automobiles usually carry onboard a tray of fifty 12-volt lead-acid batteries, adding up to 1,200 pounds of additional weight.

Recharging these batteries is in itself a cumbersome process, which usually takes place at home by plugging the vehicle into the home electric grid overnight. A typical 15 KWh recharge using the standard 110-volt circuit with a 15-amp circuit breaker can take up to ten hours, since the maximum amount of energy this type of home circuit can transfer is 1.5 KWh per hour. Even using a 240-volt circuit, with a 40-amp circuit breaker, can only deliver a maximum of 9.6 KWh per hour. This arrangement cuts the charging time to less than 2 hours but still does not compare to the five minutes it takes to fill the gasoline tank of the automobile. New high-energy batteries such as NiMH and lithium-ion have provided for longer distances, up to 2-3 times the range, but at significantly increased costs; even up to ten times as much, compared to lead-acid batteries. Nevertheless, the cost of these batteries is projected to drop significantly when these batteries are mass-produced.

The cost of running an electric vehicle is cheaper than using gasoline. Current electric models consume around .3 kWh per mile per ton of weight. Assuming the cost of electricity is 10 cents per kWh, a 100-mile-long trip would cost $3. With gasoline at $3.50 per gallon and automobile fuel efficiency at an average of 25 miles per gallon, the same trip will cost $14 in a gasoline-powered automobile.

Except for their limited range and the time it takes to recharge the vehicle batteries, electric automobiles perform similarly to their gasoline-powered counterparts. They are capable of acceleration performance, which equals or exceeds combustion engine performance. Due to the fact that electric vehicles can have multiple electric motors connected directly to each wheel, they can use all the wheels for acceleration or braking, thus increasing the overall traction and efficiency.

Recently the battery electric vehicles have been revisited and a few pilot programs have started. For reasons unknown there is resistance by the U.S. automakers to produce these automobiles. The large three automakers, Ford, General Motors and Chrysler, all produced several models in the 1990s but have since scrapped most of these programs to develop, produce and market electric hybrids. The big three automobile manufacturers also claim that lack of customer interest is the primary reason these vehicles are not produced. In other words, cheap gasoline is available on every corner and for that reason Americans do not much care for a battery-powered automobile today.

Hybrid Electric Vehicle (HEV)

In a hybrid electric vehicle, the internal combustion engine, used as the primary source of power, is combined with an electric motor. They are differentiated by how the two individual parts interconnect to power the vehicle. Regardless of the type, they use regenerative braking to recover the power. Regenerative braking means that the electric motor runs backward when braking and slowing down, in which the electric motor acts as a generator charging the onboard batteries, thus recovering the braking power. This feature is one of the components responsible for the increased overall vehicle efficiency. The two major hybrid designs are parallel hybrid design and series hybrid design.

In parallel design, both the combustion engine and the electric motor are connected to the vehicle's transmission. Most of these hybrid designs utilize the primary gasoline engine for highway driving. As the car moves, a generator continuously charges the onboard batteries. These batteries provide power to the electric motor, which is directly connected to the vehicle's wheels, and during climbing and acceleration the electric motor provides additional power.

In the series design, the primary internal combustion engine is not connected to the vehicle's transmission. The gasoline engine provides power to a generator. Electricity produced by the generator runs the electric motor, which powers the vehicle, bypassing the batteries completely.

The generator also produces electricity to charge the onboard batteries. When a large amount of power is needed, both the gasoline engine and the batteries provide power to the electric motor.

Though parallel design is the preferred system of automakers today, both designs utilize the gasoline engine as a primary source of power. These vehicles cannot run on electric power alone and so they are sometimes referred to as "assist hybrids." Both concepts, nevertheless, save a considerable amount of fuel. Hybrid design automobiles typically increase the fuel efficiency by 50 percent. That, however, is its limitation. The gasoline use cannot be changed and is fixed for the life of the automobile, thus to simplify the explanation, hybrids are primarily gasoline engines with increased fuel efficiency. The Honda Insight uses this design.

Full hybrid electric vehicle design is a vehicle that can run on electric battery power alone or the internal combustion engine power alone, or it can run on a combination of both. This system utilizes a higher capacity battery pack, for battery-only operation. An onboard computer controls both parts of the system and determines which part of the system should be running. The computer shuts off the internal combustion engine when the electric motor, powered by the batteries, is a sufficient source of energy. The Toyota Prius and Ford Escape models utilize this design.

Plug-in Hybrid Electric Vehicle (PHEV)

A plug-in electric hybrid is the "Cadillac" of hybrids. A plug-in hybrid electric vehicle adds an additional feature to the full hybrid electric vehicles. It provides a plug-in capability, where the onboard batteries can be recharged from the home electric power grid, thus making it even more versatile. A plug-in hybrid electric vehicle can run on battery power alone just as an electric vehicle with the gasoline engine turned off. Likewise, when the batteries are exhausted it can switch to a stand-alone hybrid electric vehicle to run on a gasoline mode only. Additionally, these hybrids can be multi-fueled, where the gasoline can be substituted by diesel, biodiesel or ethanol. At the end of the day these vehicles can be plugged into the home electric grid to recharge overnight and begin the new day with fully charged batteries.

The versatility of the plug-in hybrid electric vehicles is what makes it a perfect choice to begin the fuel transformation. The plug-in hybrid electric vehicle is the only design in which customers can utilize the cheapest fuel available at will. When electric power is cheaper compared to gasoline, the consumer conscious to save on the cost of fuel will make the most of the electric power. Thus, the cost of fuel and the cost of driving are in the customer's hands. The customer at last has a choice.

That is as efficient as it can get! The plug-in hybrid vehicle has provided a solution to the battery inefficient storage of electric power. With the ability of plug in technology, the customer has a choice to use the inexpensive electric power for most of the daily chores, while using the gasoline for extended trips only. Yet for some reason, the plug-in hybrids are not made by the big automakers. Individuals themselves have to experiment with converting existing hybrids to plug-in hybrids by adding additional batteries and a grid recharging capability.

Electric Solution (The Brazil Way)

Brazil sets an excellent example of the potential of dual fueled vehicles. Brazil currently replaces 40 percent of its gasoline demand with ethanol, which it produces from sugar cane. Sugar cane is easily grown and abundant to Brazil. In the 1990s Brazil introduced flexible-fuel vehicles, which run on pure ethanol or a blend of ethanol mixed with gas (20 to 25 percent ethanol, 75 to 80 percent gasoline). As of 2005, new flexible-fuel cars in Brazil have outsold gasoline cars. Plans are that within another two years, 90 percent of all new cars will be sold as flexible-fuel cars. The United States, on the other hand, does not have the ability to produce large quantities of ethanol; however, this idea can easily be adopted in the U.S. by replacing the ethanol with electric power.

Equipping the entire American automobile fleet with the capability of plug-in technology would allow Americans to utilize grid electric power when electric fuel is cheaper than gasoline. A fleet of plug-in hybrid vehicles would provide Americans with an inexpensive and domestically produced alternative energy, and when the world unrest causes oil prices to spike up drastically, Americans would have the ability to continue to drive within exactly the same fuel budget. There would be no need to panic. When the cost of gasoline reaches a level, which Americans are unwilling to cross, most Americans would just simply utilize electric power. For the most part, the price of

Electric Solution (The Brazil Way)

gasoline would become irrelevant as the cost of driving remained unchanged. In fact, with this option Americans are given an economic choice. They can pay the increased price of gasoline, if that is more convenient, or they can utilize the electric grid.

The time required for this type of transformation should be no more than fifteen years, or twenty years if the government felt that it was necessary to give a long warning to the automotive industry. The first step would require Congress to enact a law that as of 2010, for example, only new plug-in electric hybrids could be sold in the U.S. Following that decision, within the next fifteen years most of the old transportation fleet would be scrapped. The old models would slowly be replaced with the new plug-in electric hybrid models. By 2025 the gasoline-only automobile could be an antique.

To make it easier for the automakers, these automobiles need not include large battery packs when the automobiles are sold. These can be added later. What should be provided, however, are the grid plug-in capability and the space to add the battery pack at a later date. They can be added when gasoline gets expensive and the low-priced electric power begins to be appreciated. The design of adding batteries should provide for simplicity, in which the customers can easily install newly purchased sets of batteries on their own and be ready to go. Customers should be able to purchase and install just a small number of batteries, say 4 to 6, to cover just a few miles on electric power daily, or they should be able to purchase a full set of 30 to 50 batteries to achieve maximum distance on electric power only.

Most automobiles drive less than 20 miles per day, with longer trips only on occasion. Thus there is no need to carry an onboard battery pack to drive long distances. Just a small set of batteries, one dozen for example, could provide the consumers with a sufficient amount for most of their driving needs, provided they plugged the batteries into an electric grid daily, similarly to what most Americans do with their cell phones now. Just as with plugging in the cell phones, though it is a cumbersome daily chore, it is not such a big deal. Thus the battery storage problem as it pertains to an automobile has been resolved; actually has been resolved for quite a few years.

Assuming an average electric automobile uses .3 kWh per mile, then a fleet of 200 million automobiles on the U.S. streets today driving 10,000 miles per year would consume 600 billion kWh. According to the Energy Information Administration, the United States consumed 3,656 billion kWh in 2003. The nameplate capacity of these power plants is over 8,000 billion kWh, establishing electric utilization of approximately 45 percent, in which the need for more power during the peak hours accounts for the oversized capacity available. Large amounts of power are available during the off-peak hours and if charging of batteries was done during the off-peak hours, then the electric companies would be well positioned and capable of providing enough power to drive the electric plug-in hybrid automobiles now.

Even with an estimate of 20,000 miles per year, the electricity needed to run 200 million automobiles would be no more than 1,200 billion kWh, which is still within the limits of available power. (The .3kWh per mile considers 2,000 pound or smaller vehicles, a category, which only a small number of foreign models fall into today. Most U.S. models, including the large 8,000 pound SUVs, vans and trucks, need considerably more power.)

To visualize the results of the plug-in hybrids at work and what effect they could have on the U.S. economy, let's go through a hypothetical example. The question to address is this: Provided all vehicles have plug-in capability, how high could the price of gasoline go before Americans experience a significant fuel cost increase? The parameters of this exercise are as follows:

- Every automobile in the United States has a plug-in capability, capable of running on electric power only, gasoline power only or a combination of both.
- Electric costs remain the same (10 cents per kilowatt-hour), while gasoline prices fluctuate with the market.
- The electric vehicle has a battery pack capable of driving a minimum of 20 miles per day and it uses .3 KWh per mile.
- The typical average daily commute is 20 miles and Americans drive their vehicles 10,000 miles per year on average.
- The driver has the option to plug in the vehicle daily to take full advantage of the inexpensive electricity.
- The new transportation fleet has a fuel efficiency of 50 mpg, which is double that of today's requirement of 25 mpg for light-duty vehicle fleet. (Keep in mind that many current models already significantly exceed 50 mpg.)
- This will be compared to the average of the light-duty vehicle fuel efficiency of 20 mpg on the U.S. streets today. (According to Energy Information

Administration Annual Energy Outlook 2006, the estimates of average fuel efficiency figures are 20.2 for 2004 and 20.4 for 2010.) So 20 mpg it is.

With the gasoline at $3.50 per gallon, the average cost per mile today is:

$3.50 / 20 = $.18, or 18 cents per mile, regardless of the amount of miles driven each year. With 10,000 miles driven per year, the yearly budget would be $1,800. With gasoline-only vehicles, the cost of driving each mile automatically increases proportionately to gasoline prices. Thus, it will cost 36 cents per mile with $7 gasoline, 72 cents per mile with $15 gasoline, and $1.44 per mile when the gasoline reaches $30. With gasoline-only vehicles Americans have only one choice, either pay or don't drive.

Hybrid technology alone would produce very promising results. A fleet of hybrid vehicles having 50 mpg average fuel efficiency would spend 18 cents per mile if gasoline prices reached $9.00 per gallon. And with the frequently mentioned 80 mpg target; the price of gasoline could reach $14.40 per gallon for Americans to continue to drive on the 18 cents per mile, a budget of 2006, which Americans are currently accustomed to.

Plug-in hybrids compare even better. If the average automobile travels 10,000 miles per year, and 20 miles was the typical daily commute, then during the year the vehicle will travel 7,300 miles (365 x 20) within the 20 mile threshold and 2,700 miles exceeding the 20 mile range. In this example the driver can utilize the electric grid for 7,300 miles and be forced to drive using gasoline for the balance of 2,700 miles. Of course nothing is ever this exact, but for the purpose of example let's assume that it is.

Electric Solution (The Brazil Way)

With gasoline at $3.50 per gallon, the cost per mile driven is:

Cost of gasoline miles; 2,700 miles / 50 mpg x $3.50 per mile = $189

Cost of electric miles; 7,300 miles x .3 KWh per mile x $.10 per KWh = $219

Total cost per 10,000 miles driven is 4 cents per mile = $408

With gasoline at $10 per gallon, the cost per mile is 7.6 cents. In fact, the gasoline price would need to reach $30 per gallon for the cost per mile to reach 18 cents.

If we were to look at the plug-in hybrid electric vehicle another way, from the point of view of how many miles this type of vehicle would travel on one gallon of gasoline, an automobile with a fuel efficiency of 25 miles per gallon driving 25 miles on gasoline and 75 miles on inexpensive electric power, for every 100 miles driven, would have a fuel efficiency of 100 miles per gallon of gasoline. Applying the same ratios, President Clinton's 80 miles per gallon proposal could be viewed as an automobile having a fuel efficiency of 320 miles per gallon.

This explains the need for government intervention. None of the U.S. automakers would ever seriously promote the dual-fueled automobile idea on their own, especially when combined with high gasoline fuel efficiency. Fuel efficiencies of 100 to 300 miles per gallon would leave the U.S. oil companies and OPEC with a lot of unsold oil. Drastically reduced oil demand would cause the price of oil to plummet, and with tumbling oil prices and cheap gasoline U.S. consumers would once again demand large, fuel-inefficient automobiles. Thus, the U.S. automakers, in order to stay in business, would be forced to go back to the old ways of producing what the public demands—the large gas-guzzlers.

Therefore, if the government established a permanent mandate of dual-fuel vehicles, allowing only the plug-in electric hybrid vehicles to be sold in the United States, the U.S. auto industry could begin producing fuel-efficient automobiles without any negative repercussions. Americans would certainly learn to live with the inexpensive gasoline and OPEC, including the U.S. oil industry, would simply need to adjust to cheaper oil prices.

What about those who drive 20,000 or 30,000 miles per year? Increasing the size of a battery pack capable of supplying enough power to drive 40 or 60 miles per day before it is completely discharged would duplicate the numbers given above. Thus in the future, when purchasing a new vehicle, the customers could simply estimate how many miles they drive and then buy the appropriate battery pack. If the condition changes, they could simply add a few batteries to accommodate their new demand.

Unfortunately the existing battery limitation does not cover long trips, family vacations and business trips. These trips will continue to use gasoline and the price of gasoline will dictate the length and the frequency of these trips. Utilizing the most modern hybrids, the 80 miles per gallon models, these trips will continue to cost the same until the price of gasoline reaches $15 per gallon. With a little luck, more efficient electric storage media will be developed long before the price of gasoline reaches these levels.

Electric Solution (The Brazil Way)   169

    The transformation of the U.S. transportation fleet from internal combustion vehicles to plug-in electric hybrid technology would certainly create new jobs. In addition to an improved smaller internal combustion engine, the new vehicles would need many new additional electrical components, including a sufficiently large generator to provide power to the newly added electric motor, or several motors if one is provided at each individual wheel. Specialized computers will be needed to run the whole system efficiently. And most importantly, the need for a new large battery pack will evolve into a new booming battery industry. No longer would cars require just one or two batteries to start the engine and to run the onboard electric system, the lights, power locks, power windows, power seats, sun roofs and the radios; each car would now require as many as ten batteries to power the vehicle itself. With such a large increase in demand, new businesses would form fairly quickly and this increased interest could eventually produce new ingenious technologies in the way batteries store power. This could, one day, even lead to a design where a single battery could store enough power to replace ten, fifty, perhaps even hundreds of the batteries of today. This kind of battery capacity could bring the price of a barrel of crude oil back to the single digits, possibly eliminating the fuel we use in transportation altogether.

    To conclude, addressing the transportation sector only, the transformation of the light transportation fleet to the plug-in electric hybrid vehicles would not achieve complete energy independence by itself, but it would certainly pave the way. It is likely that the U.S. would continue to import oil from around the world. There would, however, be a significant difference. When the price of oil doubles, triples, or even increases by a factor of ten, similar to what Americans have become used to in the past few years, American families would no longer need to panic; rather, they would be able to utilize the abundant inexpensive alternative fuel, electric power, making the price of gasoline irrelevant and obsolete. Americans would at last be able to, politely but resolutely, graciously but respectfully, using a few choice words, tell OPEC what they can do with their oil if they are unwilling to sell it at a reasonable profit. Finally, there would no longer be any need to protect the flow of oil due to fear of supply interruptions. No more killing, no more massacres; American soldiers could finally come home.

For detailed information on electricity go to:

http://www.eia.doe.gov/kids/energyfacts/sources/electricity.htm
http://en.wikipedia.org/wiki/Electricity
http://en.wikipedia.org/wiki/Electric_vehicle
http://www.greenhybrid.com/shop/
http://en.wikipedia.org/wiki/Hybrid_vehicle
http://www.eia.doe.gov/fuelelectric.html
http://en.wikipedia.org/wiki/Lead_dioxide

7.
Conservation Solution (The U.S. Way)

Conservation is another handy, effective and very powerful tool. Several different conservation options will be introduced here. The first conservation option will just expand on a program the U.S. government has been using since the mid-1970s, the "CAFÉ Standards." The existing CAFÉ standards will be examined and then the program will be modified to see if a somewhat altered approach could provide better results. Because the CAFÉ program was designed for interaction between the government and the auto industry, excluding the customer; a second version, a varied version, the "License Plates Fix" was added, in which an American driver can interact with the auto industry directly.

CAFÉ (The Corporate Average Fuel Economy Program)

Of the 20 million barrels of oil per day that the U.S. currently consumes, 14 million barrels of it has been used in the transportation industry. Of that, according to EIA Energy Outlook 2006, the light-duty fleet of cars used 8.51 million barrels per day in 2004. Personal vehicles include vans, SUVs and light-duty pickups. The CAFÉ program was specifically designed to deal with this sector of transportation.

Recognizing the need to conserve energy, Congress passed the Energy Policy Conservation Act; on December 22, 1975, it was enacted into law. This act added a title V "Improving Automotive Efficiency," to the Motor Vehicle Information and Cost Savings Act, and established corporate average fuel economy requirements for new passenger cars and light trucks sold in the United States. The Arab oil embargo of 1973 was directly responsible for the creation of the act. The main purpose of the act was to reduce the nation's dependence on foreign energy as it focused on the substantial energy consumption of the transportation sector.

Corporate average fuel economy (CAFÉ) is the production weighted average fuel economy of the manufacturer's fleet of new cars and light trucks. CAFÉ standards set minimum requirements of an average number of miles a vehicle travels per gallon of gasoline. Individual vehicles are not required to meet the minimum fuel efficiency but individual manufacturers must meet the minimum fuel economy by averaging fuel economy of all of their vehicles manufactured in a given year.

Authority to establish the CAFÉ standard was assigned to the National Highway Traffic Safety

Administration (NHTSA) by the Secretary of Transportation. The National Highway Traffic Safety Administration is responsible for establishing the CAFÉ standards. CAFÉ standards have been established with the goal of doubling the fuel economy by 1985. Two separate CAFÉ standards were established; one for passenger cars and another one for light trucks.

Passenger car fuel economy standards had been established to be 27.5 miles per gallon by the year 1985. Intermediate goals were established to remain on schedule. The first standard of 18 miles per gallon (mpg) had been set for 1978. Standards of 19 mpg in 1979, 20 mpg in 1980, 22 mpg in 1981, 24 mpg in 1982, 26 mpg in 1983, 27 mpg in 1984 and finally 27.5 mpg in 1985 were also documented.

Light truck fuel economy standards were established in a similar manner. The first fuel standards for light trucks had been set for 1979. For the first three years, from 1979 to 1981, the standards had been established for two-wheel drive and four-wheel drive vehicles separately. Starting in 1982, a combined standard had been established with the two-wheel and four-wheel standards as optional. Beginning in 1991, the optional standards were dropped and the combined standard became the rule.

The fuel economy standards for light trucks were established at 17.2 mpg for a two-wheel drive and 15.8 mpg for a four-wheel drive in 1979. The combined 17.5 mpg standard was established in 1982. In 1983, the standards were raised to 19 mpg and in 1985 the standard reached 19.5 mpg. Light trucks are trucks with a gross vehicle weight rating of 8,500 pounds or less. This was increased from the original gross weight of 6,000 pounds in 1980. Light trucks that weigh more than 8,500 pounds do not have to comply with the fuel economy standards. The light truck category includes pickups, vans and SUVs.

Overview of CAFÉ standards
Courtesy of Department of Transportation

MODEL YEAR	CAFE STANDARDS			
	PASSENGER CARS	LIGHT TRUCKS		
		COMBINED	2 WD	4 WD
1978	18.0			
1979	19.0		17.2	15.8
1980	20.0		16.0	14.0
1981	22.0		16.7	15.0
1982	24.0	17.5	18.0	16.0
1983	26.0	19.0	19.5	17.5
1984	27.0	20.0	20.3	18.5
1985	27.5	19.5	19.7	18.9
1986	26.0	20.0	20.5	19.5
1987	26.0	20.5	21.0	19.5
1988	26.0	20.5	21.0	19.5
1989	26.5	20.5	21.5	19.0
1990	27.5	20.0	20.5	19.0
1991	27.5	20.2	20.7	19.1
1992	27.5	20.2		
1993	27.5	20.4		
1994	27.5	20.5		
1995	27.5	20.6		
1996	27.5	20.7		
1997	27.5	20.7		
1998	27.5	20.7		
1999	27.5	20.7		
2000	27.5	20.7		
2001	27.5	20.7		
2002	27.5	20.7		
2003	27.5	20.7		
2004	27.5	20.7		
2005	27.5	21.0		
2006	27.5	21.6		
2007	27.5	22.2		

In 1985, Congress gave the Department of National Highway Traffic Safety Administration the authority to set higher or lower CAFÉ standards. From 1986 to 1990 the passenger car standards were lowered to 26 mpg. In 1990 the passenger standards were raised back to 27.5 mpg and the standard has remained unchanged ever since.

The practice of the agency was to establish new standards for one or two years in advance. On November 15, 1995, Congress enacted the Appropriations Act for 1996, eliminating all funds to prepare and prescribe corporate average fuel economy standards for passenger automobiles and light trucks. Pursuant to that act, CAFÉ standards remained frozen at 27.5 mpg for automobiles and 20.7 mpg for light trucks. No comments were available as to when the funds could be reinstated for the work to continue.

On October 23, 2000, the Appropriations Act for the year 2001 was enacted. While not releasing any new funds to prepare and propose new CAFÉ standards, the Appropriation Act had decided to fund a study by the National Academy of Sciences (NAS) to evaluate the effectiveness and impact of CAFÉ standards. The NAS report was submitted to Congress on July 30, 2001. The report concluded that technologies exist that could significantly increase fuel economy. The report concluded that the implementation of new economy standards would require trade-offs, between environmental benefits, safety and cost of vehicles, energy independence and customer preference.

The provision that restricted the implementation of new CAFÉ standards was finally eliminated in the fiscal year of 2002. Due to a lack of information available, the agency determined to leave the 2004 standard unchanged at 20.7 mpg. On December 16, 2002 NHTSA proposed new CAFÉ standards for light trucks to be enacted in the years 2005 to 2007. This proposal established the standards at 21 mpg for 2005, 21.6 mpg for 2006 and 22.2 mpg for 2007. The proposed standards established the maximum feasible fuel economy levels manufacturers can achieve and relied on the data from the NAS fuel economy report.

The program considered four separate factors to determine the Maximum Feasible Fuel Economy Levels:

1. Technological feasibility.
2. Economic practicability.
3. Effect of other standards on fuel economy.
4. Need of a nation to conserve energy.

Technological Feasibility - NHTSA has relied heavily on the information provided by General Motors, Ford and DaimlerChrysler to establish these levels. On December 16, 2002, the agency published a Notice of Proposed Rulemaking, proposing to increase the CAFÉ standard for light trucks from the current level of 20.7 mpg to 21 mpg in 2005, 21.6 mpg in 2006 and 22.2 mpg in 2007. GM suggested the levels be at 20.4 mpg in 2005 and 2006, and at 20.6 in 2007. Ford had agreed with the NHTSA proposal. DaimlerChrysler announced that it would be able to achieve levels of 21.3 mpg in 2005, 21.6 in 2006 and 22.2 in 2007. This information seemed to suggest that the automaker's recommendations were just simply averaged out and recommended by the NHTSA.

Economic Practicability - Economic practicability considerations are numerous. The cost incurred by car manufacturers to meet the increased standards is being compared to cost savings, which benefit society. The report estimated that the cost to meet the new standards per car would be $22 in 2005, $67 in 2006 and $106 in 2007. The total cost would be $170 million in 2005, $537 million in 2006 and $862 million in 2007. Arguments have been put forth that the increased standards will restrict the availability of large engines which would lead to job losses. The recreational industry has argued that the new standards will indirectly impact the sales of trailers and recreational vehicles by reducing the towing capacities of the new light truck fleet. The report estimated benefits to society as a result of the increased CAFÉ standards to be $29 in 2005, $83 in 2006 and $121 in 2007. The total value of these benefits was estimated to be $218 million in 2005, $645 million in 2006 and $955 million in 2007. The cost savings factors include greenhouse emissions, economic cost to import petroleum into the U.S., consumer choice, gain or loss of jobs and military presence to maintain the U.S. presence to guard foreign oil in unstable regions of the world. Interestingly, the study showed that increased fuel efficiency would lead to increased employment. This shows that as the standards increase, employment would increase proportionately.

<u>The Effect of Other Federal Vehicle Standards on Fuel Economy</u> - This section analyzes the Federal Vehicle Safety Standards and how they impact vehicle weight, where additional weight may cause a reduction in fuel efficiency. It considers all mandated and voluntary safety requirements. Some of these include occupant crash protection, child restraints and improvements to tire standards. Weight reduction of an automobile has been

directly associated with increased death rates. The big objection from the safety standards point of view was that if the manufacturer, who needs to increase the fuel efficiency, will just simply make the vehicle lighter, by making a lighter body frame and using lighter metal; then when the vehicle is involved in an accident it is more vulnerable.

<u>The Need of the Nation to Conserve Energy</u> – The Energy Policy and Conservation Act has been specific. The department was to balance the need to conserve energy with the technological and economical problems and provide us with some solution that will help us to achieve oil independence in our future. Congress directed that all factors of the research must be taken into consideration before any recommendation is made. NHTSA has been criticized by many groups, which claim that basing the CAFÉ standards on the manufacturer's information limits the agency's decision-making. They argue that advanced technologies available to manufacturers to increase fuel efficiency will not be utilized.

Reading the extensive Department of Transportation report on CAFÉ, one may conclude that their research had set out to please everyone, which included the party in power, Congress, the public, the automakers, the oil industry, the safety department and even the environmentalists, including all of the special interest groups. So except for the groups promoting much higher standards, everyone was pleased at the end of this process.

When a manufacturer fails to meet the established standards he faces civil penalties. Noncompliance penalties are established by the NHTSA. The current penalty is $5.50 for every one-tenth of a mile per gallon over the standard. To simplify the math, it is $55 per mile for each vehicle. If a manufacturer's CAFÉ was one mile below the established standard, then the penalty would be $55 for each car in a fleet. What a joke! With an average automobile cost of $20,000 dollars, $55 is not much of a penalty. Even with a default as high as ten miles below the average standard, the penalty would still only take $550 per car from the automaker's pocket. Civil penalties may not be the best way to produce results but having insignificant penalties, like the ones that exist today, is a slap in the face of the public paying the bills.

During the origins of the Energy Conservation Policy Act, the purpose was well-defined and the job was simple. The purpose was to increase fuel efficiency by means of forcing the automotive industry to build more efficient cars. A ten year effort had been put forth to double automobile fuel efficiency. Intermediate goals were established and enforced, and the program worked. The program did everything it set out to do. The automobile fuel efficiency had increased, doubled in this case, and fuel consumption had decreased. During the CAFÉ implementation from 1978 to 1985, fuel consumption on the U.S. highways had decreased by some 10 percent. There is no doubt that part of this decrease was caused by the fuel price increases between 1979 and 1980, but CAFE standards also contributed significantly.

What then went wrong? By 1985 the Energy Conservation Policy Act had been buried in bureaucracy. Politics had taken over and all progress had come to an abrupt halt. In 1986 the conservation requirements were reversed. They were reduced to 26 mpg. The table below explains it all. From 1975 to 1985 the fuel economy of automaker's fleets had doubled. From 1985 to 2006 the standard remained exactly the same. It would seem that Congress had the right idea in 1995 when it cut the funding to implement the CAFÉ standard. It seemed that they too had figured out NHTSA's uninspiring performance. Then a new administration came to power and the program was reborn. NHTSA managed to increase the fuel efficiency by 1.5 mpg in the following three years. What was Congress thinking?

To summarize; the organization, the National Highway Traffic Safety Administration, had forgotten the purpose for which it was originally created and had failed sadly to accomplish the intended goal. Had the project continued on for another 20 years the way it was designed, doubling the fuel efficiency every ten years, the fuel efficiency of the U.S. automobile fleet would have gone up. It would have increased from 27.5 mpg in 1985 to 55 mpg by the end of 1995, and to 110 mpg by the end of 2005.

It should be noted that CAFÉ deals with new vehicles only. The transportation fleet term used means the fleet of newly produced cars for that specific year. The fuel efficiency of the U.S transportation fleet of all light-duty vehicles on the road today varies significantly. It will take 20 years to replace the U.S. auto fleet. Therefore the 22.7 mpg standard requirement proposed for 2007 applies only to one-twentieth of the total U.S. fleet. Provided no increases were made in future years, the U.S. fleet would accomplish the 22.7 mpg standard in 20 years.

Now that the existing program is familiar and the shortfalls have been identified, could a somewhat altered version be adopted to take the U.S. away from oil dependency? The government would have to enact a program which would set guidelines that would produce the desired result within a specific time frame. The government would also have to establish intermediate goals to measure the progress and would have to introduce stiff penalties for not meeting the standards.

Increase Fuel Efficiency by Modifying CAFÉ

One of the major pitfalls of the government program was only establishing the fuel standards for one or two years in advance. This approach has proven detrimental to CAFÉ's success. Had a 20-year plan been adopted, then the results may have proven much different. The industry complaint has continuously been that not enough lead time is provided to the auto manufactures. Automakers claim that with sufficient lead time, new advanced technologies could be employed. Let's examine a hypothetical example with longer duration and a longer lead time.

Automobile producers would again be required to meet the average fuel economy standards. To simplify the math, let's first eliminate the two separate groups, the automobiles and light trucks, and establish only one standard. The weighted average fuel economy of any manufacturer's fleet of cars and light trucks could be established similarly to the existing CAFÉ standards, by taking all of the automobiles, trucks, vans and SUVs produced and averaging out their fuel efficiencies. We can call this the manufacturer's CAFÉ. Thus, if a manufacturer's fleet is comprised of 100,000 automobiles with 20 mpg capabilities, 100,000 automobiles with 40 mpg capabilities, and 100,000 automobiles with 60 mpg capabilities, then the manufacturer's CAFÉ would be 40 mpg.

The compliance of Detroit and all the foreign autoproducers would need to be revisited too. The penalties that currently apply are meaningless. The irrelevant penalties in existence today, such as the $55 per mile per car, do not discourage anyone. The manufacturer will gladly pay a low penalty if it means a higher bottom line. Stiff penalties of $1,000 to $5,000 per mile per car would be a nice incentive to all automakers to do proper accounting of what they put up for sale.

The program is simplicity itself. It would adopt an act to increase the fuel efficiency of the CAFÉ standards by 10 percent per year. That's it! Beginning with an average fuel efficiency of 20 miles per gallon and increasing the fuel efficiency by 10 percent annually; the fuel efficiency would double to 40 miles per gallon in seven years, to 80 miles per gallon in fourteen years and to 160 miles per gallon in twenty-one years. It is worth noting that as of 2006, 80 miles per gallon vehicles already exist, though they are few and far in between. What about 160 miles per gallon? Well, maybe in twenty years.

This hypothetical example uses 10 percent per year, just to make the illustration simple, but be sure that 5 percent, 7 percent or even 15 percent would be just as valid. There are many other derivatives that could provide similar results. A program that increases the average fuel efficiency by three miles per year would increase the fleet's fuel efficiency from 20 to 50 miles per gallon in ten years and to 80 miles per gallon in twenty years. Another variable worth attention would be establishing a minimum fuel efficiency requirement. It is just a matter of math to select an appropriate set of figures which would best fit the U.S. economy. Anyone can provide many other feasible alternatives. To sum up, whatever the U.S. Congress can agree to would be better than doing nothing.

During the early years of this newly established program, the production of alternative fuel automobiles may not be too lucrative an option and may not resonate within the automobile industry. However, when the required manufacturers average fuel efficiency starts to approach 50 mpg or so, and the manufacturer is required to produce a 75 mpg automobile for every 25 mpg gas SUV, the auto manufacturers will begin to use their ingenuity to meet these requirements. No one knows exactly how this type of change would affect the automotive industry and Americans, but a reasonable prediction could be made. Perhaps the following may come to pass:

During the first seven years, the fuel efficiency requirement would be increased to 40 miles per gallon. The easiest way for an automaker to accomplish a higher overall fuel efficiency, in order to avoid stiff penalties, would be to reduce the number of big gas-guzzling automobiles like some of the pickups, vans and SUVs. With the gasoline prices remaining low and the demand for large vehicles increasing, the supply and demand rule would dictate the price of these vehicles. As the supply of large vehicles begins to shrink, their price would begin to increase. Eventually the price of a large SUV would no longer justify owning one and smaller automobiles would be substituted.

With a requirement of 80 mpg (fourteen years from implementation), making a smaller number of large automobiles would only take the automaker so far. When this option was exhausted, and it certainly would be at the 80 mpg level, then to meet the fuel efficiency requirements, high fuel efficiency hybrids would need to be introduced. Automakers would start experimenting with and improving hybrid automobiles. Because of the restrictions, not too many gas-guzzling automobiles would continue to be made. Their high price would also make them undesirable to most Americans.

Adding a twist, a preferential treatment if you will, would be to contribute positively to the average fuel requirement for non-gasoline automobiles. For example, when the fuel efficiency demands 80 mpg, then the non-gasoline automobile counts as 160 mpg. Contributing positively to the average fuel requirement could spur extensive research and development of alternative fuels by the auto industry. This added twist could produce a fleet of automobiles where half run on gasoline averaging 50 to 60 mpg and the other half runs on alternative energies such as natural gas, ethanol, electric or hydrogen.

Another twist would be to limit the amount of gasoline automobiles being produced. Instead of a fuel efficiency program, the government could establish a required percentage of the automobile fleet that runs on alternative sources of energy. Say 3 percent, for example. The first year, 3 percent of the entire auto fleet would be required to run on alternative fuel, the second year this would raise to 6 percent, and then 9 percent and so on. In seventeen years, half of the fleet would run on gasoline and half on alternative fuel; and if penalties were introduced, such as $10,000 per car that is produced in excess of the allowed ratio, automakers would keep detailed inventories. In this example, the automotive fleet would need to be separated into several different categories which could include, cars, trucks, vans and SUVs; with additional weight classes of light, medium and large in each category. This would assure that the automakers would not just eliminate the need for gasoline in the smallest vehicles but would do so equally across the whole spectrum of the vehicle fleet.

To sum up, a serious conservation program would reduce the U.S. fuel consumption considerably. Applying CAFÉ standards to a light transportation vehicle fleet only may not be sufficient enough to gain complete oil independence; however, even by itself it would eliminate a significant portion.

License Plates Solution

Finally, a somewhat different approach would be a program that penalizes car owners for driving gas-guzzlers and at the same time earns substantial sums of money for the government. Every automobile in the United States has to buy license plates annually. Today the cost of license plates is very inexpensive, yet even today there are two classifications: personal automobiles and business trucks. The program described below just expands on this principle. This plan differs from the previous examples in that previously the responsibility to increase efficiency was assigned to the automaker by the government, and the automaker was responsible to carry out the task. In this example there is no responsibility assigned to anyone.

The cost of license plates would be directly linked to the fuel efficiency of the automobile. Automobiles would be classified into categories according to their fuel efficiency and each category would have its own license plates fee. The fee that the consumer pays would depend on which category the customer's vehicle falls into.

Again, let's go through a simplified hypothetical example to illustrate this type of plan in action. A category list could be devised similarly to a progressive scale like this one:

Cost of license plates:

1st through 5th year
1. Alternate energy fuel automobiles pay nothing!
2. Automobiles with fuel efficiency above 40 mpg pay nothing.
3. Automobiles with fuel efficiency between 20 and 40 mpg pay $500.
4. Automobiles with fuel efficiency under 20 mpg pay $1,000.

6th through 10th year
1. Alternate energy fuel automobiles pay nothing!
2. Automobiles with fuel efficiency above 60 mpg pay nothing.
3. Automobiles with fuel efficiency between 40 and 60 mpg pay $500.
4. Automobiles with fuel efficiency between 20 and 40 mpg pay $1,000.

5. Automobiles with fuel efficiency under 20 mpg pay $10,000.

11th through 15th year
   1. Alternate energy fuel automobiles pay nothing!
   2. Automobiles with fuel efficiency above 80 mpg pay nothing.
   3. Automobiles with fuel efficiency between 60 and 80 mpg pay $500.
   4. Automobiles with fuel efficiency between 40 and 60 mpg pay $1,000.
   5. Automobiles with fuel efficiency between 20 and 40 mpg pay $10,000.
   6. Automobiles with fuel efficiency under 20 mpg pay $20,000.

The list could continue but the illustration is clear. A cost schedule similar to the above, when established, could be posted in every License Bureau and allow the drivers to decide what they want to drive. Whatever the name, be it an incentive to Americans who drive small, high fuel-efficient cars or a deterrent to Americans who drive large gas-guzzlers, the fuel efficiency would go up. American drivers are not dumb. They would strive not to pay anything for the license plates by planning ahead. They may pay the $500 cost, even the $1,000 cost, but they will likely trade their vehicle to a more efficient automobile before they would pay $10,000 or $20,000 annually.

Consumers will have complete control of what they drive and how much they pay. In this case, selection and purchase of a new automobile would have an added variable. No longer would Americans buy an automobile just on the basis of looks and price, they would now have to decide how much they are willing to pay annually for the license plates; a new maintenance fee, if you will. A buyer would now decide how much comfort they are willing to sacrifice to pay less in maintenance fees.

When an automobile is purchased, it has a fuel efficiency assigned to it. It stays with the automobile for life. When it is resold, the new owner inherits the fuel efficiency with the car. This fuel efficiency will be used in determining the license plates fee for the life of the automobile, regardless of who owns it, thus there is no way to cheat.

The government, as in all of the previous conservation options, has plenty of leeway to choose an appropriate combination of figures. They can extend the time that is required for the final goal, raise or lower the fees, or change the fuel efficiency requirements. They can select to begin by applying the new strict fee structure to the purchase of new vehicles only, leaving the existing fleet of automobiles alone, or the choice may be to apply a stricter, progressively faster fee structure to the new automobiles, and slower, gradually increasing fee structure to the older fleet. The choices are numerous, but once again, the implementation phase of the program rests with the Congress of the United States. Once the government chooses a program and creates a similar table and posts it in the License Bureaus throughout the country, the government's job is technically finished. The only other thing left to do is to decide how to spend the large amount of revenue collected.

Award Conservation Solution

There are other types of conservation that are either already being directly or indirectly utilized, or could easily be adopted. Today, supply and demand controls the price of the utilities. As the supply shrinks, the price of the utility increases across the board, forcing the demand to slow down; then when the demand slows down, the price of the utilities lowers without regard to who actually got involved in the conservation and was responsible for lowering the price. In a sense, this is rewarding consumers who continued to waste energy just because they could afford it. What if a conservation program involved rewarding consumers that participate in conservation and punished the consumers that did not? This type of conservation could be very easily adapted to heating and cooling homes that use natural gas, electric power and heating oil.

The example that follows is not unlike a system already in existence in Florida, where the electric utilities companies already routinely reward customers willing to get involved in conservation. Cooling Florida homes in the middle of the day requires a lot of electricity. To avoid shortages during the peak hours, the electric utilities are willing to pay customers a small fee if they agree to have their air-conditioning turned off for a period of time during the extremely hot, sunny, tropical Florida days. A device, which controls the central air-conditioning and can be controlled remotely by the electric utility company, is installed at the customer's home. The homeowner hardly even notices when the air-conditioning has been turned off. The fee is small, only about $10 to $15 per month, yet even a small savings, such as this, attracts a significant number of participants. How many customers would get involved and how much comfort would they be willing to sacrifice if the utility companies offered to pay, say, "half of the electric bill?" Let's use an illustrative hypothetical example.

In this example natural gas is used, however, this system can be equally as easy with heating oil and electric power.

Natural gas companies, along with electric companies, keep a detailed record of how much natural gas a household uses each month. In the case of natural gas, the record of the past twelve months is on the monthly bill. Gas companies also provide consumers with an estimate of how much they can save should they lower the thermostat one degree. This estimate is generally 2 to 3 percent of the monthly bill. Hence, if a household lowered the thermostat from 75°F to 65°F, they could lower their bills by 20 to 30 percent. Just by structuring the price scale of the cost of natural gas to award the participating households who got involved and lowered their thermostats to save energy, substantial conservation could be accomplished.

A price structure similar to the one that follows significantly awards the participants who "put on a sweater," the famous line from Jimmy Carter. The consumer who is unwilling to participate pays the same, and the consumers who ignore all the calls to conserve and continue wasting energy would pay more. Let's see how this could work.

In early 2006, the price of natural gas peaked at around $18 per Mcf in many regions across the U.S. Let's

Conservation Solution (The US Way)

use an average price of $14 per Mcf and assume that the supply and price of the natural gas in our example remains fixed. Additionally, let's estimate that an average home uses 20 Mcf per month during the heating season, for an easy calculation. Then our example would look as follows:

The gas company would separate the utility bill into two parts. It would take the first 70 percent of last years average consumption and charge $5 per Mcf (the summer rate), and the balance of the 30 percent would be charged at $35 per Mcf. (Anyone can change the ratios any way they want and come up with a better set of numbers.) This simplified price chart would keep the gas bill exactly the same, if the consumption remained unchanged from the previous year.

If nothing was done, the bill for that month would be 20 Mcf x $14 / Mcf = $280

The do-nothing scenario would break down as follows:
First 70% of last year bill: 14 Mcf x $5 / Mcf=$70
The other 30% of last year bill: 6 Mcf x $35 / Mcf=$210
Thus the total is: $280

No change has taken place. Let's see what the gas bill would be if a customer saves 10%, 20%, and 30%.

10% customer savings:
First 70% of last year bill: 14 Mcf x $5 / Mcf = $70
The other 20% of last year bill: 4 Mcf x $35 / Mcf=$140
The total is: $210

20% customer savings:
First 70% of last year bill: 14 Mcf x $5 / Mcf = $70
The other 10% of last year bill: 2 Mcf x $35 / Mcf=$70
The total is: $140

30% customer savings
First 70% of last year bill: 14 Mcf x $5 / Mcf=$70
The total is: $70

So, 10 percent energy saved will save 25 percent of the total bill, 20 percent energy saved will save 50 percent of the total bill and 30 percent energy saved will save 75 percent of the total bill; and the natural gas shortages would be eliminated.

There is no need to see the price of natural gas fluctuate between $5 and $20 per Mcf and frightening the homeowners as to whether they can afford to live in their existing homes or whether they will be forced out. These wild price fluctuations are crazy and serve no one. Homeowners would much rather see the prices stabilize and be able to budget their expenses. Then again, maybe the natural gas suppliers love this quadrupling of natural gas prices, because that increases profits. It would be sad though, if it were done intentionally. Since the "free market" is incapable of delivering stable prices, a system like the one just discussed may be an attractive alternative.

Conservation Using Batteries

When the battery storage limitations are understood but not necessarily improved, the batteries can be applied in other situations requiring conservation. One situation in particular is to help electric utilities deal with the peak demand. Not only could the use of batteries help with the famous California rolling blackouts, which were a direct result of a shortage of electricity during the peak demand, but they could also preserve the electric grid from overloading and failing. A win–win situation could be achieved easily just by using battery storage. Suppose that the electric utility company selected an area, preferably a section of a city that was mostly residential, and provided the homeowners with batteries capable of storing a few kilowatt-hours of electricity, and installed them in such a way that they will automatically charge during the hours of least demand. Then during the several hours of peak demand, the utility company could turn off that section of the city and the households could run on the battery power alone. California could have avoided a lot of misery with a little planning.

Of course it would not have to be part of a city, forcing everyone in that section to either accept the supplied batteries or live without electricity during peak demand. Suppose that the electric utility company expects a 5 percent shortfall of electric power during the peak hours. The utility company could pay households a small fee for allowing them to install several batteries in their homes that would be charged during the night hours and automatically supply power during the peak hour, as mentioned previously with the Florida utilities. If 5 percent of households were wired in this way, there would be no need for rolling blackouts. The customers who are willing to go through the minor discomfort of storing the electric utility's batteries get paid a small fee, thus rewarded, and the utility companies can make the most of their excess electric capacity while at the same time avoiding the need for more power plants.

In the end, if anyone feels that electric utilities should not be involved at all and that the simplest and cheapest way for the electric utilities to operate is to use those rolling blackouts; the electric companies could sell the electric power during the few least demand hours at a significant discount to all the customers and let the individual customers own and control these battery systems. Customers could then supply their own batteries, charge them at night at a large discount, and utilize this cheap power during the rest of the day. In this example, however, the discounts would need to be significant to offset the cost of the batteries, since they are not cheap and need to be replaced periodically.

Finally, if this type of system was applied to charging batteries in electric–gasoline hybrid automobiles, a large fleet of plug-in electric hybrid automobiles could be used without any need for additional electric power. Electric utilities could, for example, install a device, a combination of a timer and a meter, to allow consumers to charge their vehicle batteries during certain limited night hours at a significant discount. Thus participants get to enjoy considerable savings and the electric utilities can sell the otherwise wasted "off-peak capacity."

The practice of implementing rolling blackouts is just a harsh way to control the supply and demand of available electric power. This system of power distribution does not discriminate against anyone. A certain section of a city is turned off regardless of who lives there. It is not the rich and well-to-do who lose the electric power and are not able to cool their homes for a few hours, since most of these households have backup power sources. It is the middle class and especially the elderly and the sick that lose electric power to the whole house. These people may not have any air-conditioning in the first place and rely on an electric fan to keep cool. When some of these people die because of excessive heat exposure, the weather is blamed rather then the poor planning of the electric utilities. These supposedly indiscriminate solutions, rolling blackouts, could easily become a thing of the past with a little planning and a few dollars invested.

Going back to the Florida example where the utilities pay to have the capability to turn of the home air-conditioning remotely during peak demand hours; let's suppose it became mandatory in all households across the U.S. to have the air-conditioning wired similarly. The building codes throughout the U.S. would require the air-conditioning to be wired separately. If this were to take place, then rolling blackouts would take on a little different meaning. During the high power demand times, only the air-conditioning would be turned off remotely for a couple of hours throughout the specific area. The utility companies would be capable of easily distributing the shortfall throughout the specific area equally, thus leaving the rest of the household electric power unaffected. The rolling blackouts would be limited to cooling of homes only.

Since the usage of hot weather and home air-conditioning is the culprit of larger electric demand and causes the electric shortages, the cooling should be addressed to solve the shortfalls. If the idea of giving the electric utilities the power to turn off the homeowner's air-conditioning for a couple of hours in an emergency of sort is unacceptable, then a price structure could be established as an incentive to conserve electric power by the homeowners. Staying with the previous idea of home air-conditioning having a separate meter connection, utilities could formulate a price structure to encourage serious electric power conservation.

One way could be to charge a much higher rate for air-conditioning usage during the peak hours. The three hours of highest daily demand may require a premium of double or triple the regular rate. The new rate could be a flat fee for the entire three hours every day or it could go into effect only when the temperature exceeds a certain preset temperature, such as 80°F. Another option is to have the rate fluctuate based on the temperature outside. Such as: at 80°F the price doubles and at 90°F the rate triples, and so on.

This price structure would have the same effect on conservation as discussed previously in the section on award conservation during the heating season in the northern part of the United States. Again the electric utilities could easily think up and formulate many other similar fee structures that would be acceptable to everyone.

It is unlikely, however, that the electric utilities, being monopolies, will utilize any type of conservation, be it the installation of additional capacity using batteries, some type of conservation effort of controlling the household air-conditioning, or by implementing a fee structure to increase the cost of cooling. Assuming that the electric utilities determine the rolling blackouts are unacceptable due to pressure from the consumers, the U.S. consumer is likely to see the price increase for the whole electric bill, or at best during the peak hours for the entire household, rather than just the cooling part of the bill. Therefore the poor Americans who do not use air-conditioning at all and who have been involved with conservation all along to keep their electric bills low will be required to subsidize the million dollar household cooling bills of the rich and famous.

Conservation Conclusion

Once the overwhelming power of reward conservation is understood, it can be applied in almost any application with any fuel. The energy savings of reward conservation far exceeds anything the politicians can come up with.

Do not dismiss the power of conservation! It is an efficient, effective and very powerful tool. Conservation should not be regarded as being capable of saving 5 or 10 percent; it should be regarded as being capable of saving a lot more, like 70 or 80 percent, of the energy that the U.S. uses. Consider the waste and potential savings in this: When we discuss how much energy is needed to drive a soccer mom to a game, what is really being discussed is how much energy is needed to propel an 8,500 pound SUV to get a 120 pound soccer mom to a soccer field. Enough fuel is needed, not just to propel the 120 pound mom, but also the 8,500 pound SUV. That is some waste! Lowering this weight ratio would reduce fuel consumption. How far this weight ratio could be lowered is anyone's opinion.

In conclusion, conservation can be accomplished easily everywhere. If done correctly, there is no set limit to how much can be saved. People usually resent when someone orders them to do something, like President

Carter's statement of "Put on a sweater!" This type of statement usually commands a "one word response." However, when given a choice, even a small one as demonstrated with the Florida electric utilities, Americans will get involved to a much greater extent. Next time you hear the familiar political tune of "the U.S. needs more oil, drill the ANWR, take over the Middle East, more electricity, more ethanol, more nuclear power, more, more, more, more;" consider the simpler and significantly easier course of action. Conservation!!!

For more information go to:

http://www.nhtsa.dot.gov/portal/site/nhtsa/menuitem
http://www.nhtsa.dot.gov/portal/site/nhtsa/menuitem.d0b5a45b55bfbe582f57529cdba046a0/
http://www.nhtsa.gov/cars/rules/rulings/CAFE/alternativefuels/background.htm
http://www.nhtsa.gov/cars/rules/rulings/CAFE05-07/Index.html

8.
When All Else Fails (The Europe Way)

When all else fails, use the bombshell approach of raising the price of gasoline. Up to now every option considered concentrated on preparing the U.S. economy for when or if the price of oil increases. Plug-in electric hybrids concentrate on having the ability to use the electrical power as an alternative fuel. The various modifications of CAFÉ standards, along with the license plates option, concentrated on high fuel efficiency or on replacing a part of the vehicle fleet with vehicles running on fuel other than gasoline. In all of these options, the cost of the oil itself, the cost of the gasoline and the cost of the diesel fuel powering the automobiles was completely ignored. In this section the problem will be approached from exactly the opposite point of view.

This approach will concentrate on the fuel itself and ignore everything else. Because Europe is the pioneer of this type of solution, the success in Europe will be acknowledged first. This section will examine the European approach and how it differs from the United States' approach, not politically, but only how it compares in terms of use of natural resources and how it varies from the U.S. in the transportation industry. Secondly, an attempt will be made to develop a similar plan and examine whether or not it could be adopted for use in the U.S.

When Europeans visit America, they are truly amazed how inexpensive gasoline is. The cost of gasoline in Europe is approximately three times more than what Americans pay. While Americans complain of having to pay $2.00 to $3.00 per gallon of gasoline, Europeans dish out $6.00 to $7.50 for every gallon of gasoline they use. A large part of this cost is tax, which is imposed on gasoline by the European local governments. With the large revenue derived from the gasoline fuel tax, countries such as France and Germany can fund research into alternative sources of fuel. At the moment, as of 2006, Germany is the undisputed leader in research, development and installation of solar and wind power. Furthermore, Europe is at least a decade ahead of the United States where alternative fuel automobile research is concerned. Hybrid electric automobiles, providing double the fuel efficiency, have been common on Europe's streets for quite some time now. Electric automobiles are routine, new designs are being tested, infrastructure to provide recharging in the large cities of Europe is being put in place and the technology to enable electric automobiles to travel extended distances is being perfected. America finds themselves far behind!

Compared to the upbeat European advances in electric transportation, the United States' hybrid vehicle technology, providing high fuel efficiency, exists only in foreign designs such as Hondas and Toyotas. In the U.S., Americans have yet to see an electric automobile on the showroom floor of an auto dealer. This kind of indifference and inaction is likely to lead to a U.S. downfall in the future. Europe and Japan now have technologies, which America lacks and unless a path toward alternative technology begins soon, the U.S. will find itself having to pay exorbitant sums of money to foreign governments for their technologies.

Furthermore, Europe and other foreign countries, which have far more advanced technologies, will start filling the U.S. demand for alternative fuel automobiles when the need arises; and American automakers, which already struggle financially, will find themselves going out of business. Because it will be highly unlikely that the U.S. domestic automakers will be able to make up the lost ground of technological progress, when the gasoline prices suddenly jump five or tenfold, especially when, due to these high fuel prices, the sales of large automobiles (the bread-and-butter of the U.S. domestic automakers), comes to a screeching halt. Rather than thinking of new investments to more fuel-efficient vehicles, the U.S. automakers will find themselves discussing how many employees should be laid off. Should this scenario happen suddenly, then the more technologically advanced nations of Europe, Asia and the far East will take over the U.S. automotive industry permanently.

Because of this advanced technology, European countries can acknowledge that they have collectively found a solution that can lead to oil independence. It is not a solution that everyone agrees with and it would most certainly not be the preferred solution in the U.S., because no one ever agrees with raising the price of gasoline; however, it is the most comprehensive and most complete solution out there that can do the job. That is, it will slow consumption while promoting conservation, it will encourage production of alternative energy and it will prevent the terrible consequences of shortages, described previously, all at the same time. It is designed to lead the U.S. to become truly independent of foreign oil.

What is involved is increasing the price of oil to where it is no longer a desirable commodity, and where most or all of the other options available become far more attractive. Before any proposal to increase the price of oil is to be made, some considerations have to be made; such as what impact would such a proposal have on the economy and the consumers and how would current trading partners view the United States. Having said that, let's first review the alternative options that are out there waiting impatiently to replace the imported oil.

- Oil, which is still in the ground
- United States oil shale
- Oil sands of Canada
- Liquid biofuels such as biodiesel, ethanol and methanol
- Natural Gas

- Hydrogen
- Wind and solar energies
- Electric

All of the options mentioned are ready to take over for imported oil. Unfortunately, all of these options are more expensive, therefore uneconomical, and are consequently not being seriously pursued as of now.

10% Solution

The 10% solution is very simple. It means to increase the price of gasoline by 10 percent per year. How would a 10 percent per year increase in the price of gasoline help? The 10 percent increases will double the price of gasoline every seven years. Starting with the current price of $3 per gallon, it will cost $6 per gallon in seven years and $12 per gallon in fourteen years.

A 10 percent increase in gasoline prices each year would provide Americans with a clearly defined price in future years and would allow them to plan accordingly. In 2005, the price of gasoline increased from $2 to $4 (a 100 percent increase), and back to $2.50 in a matter of one year; therefore a 10 percent per year increase would not cause too much hardship on anyone. In fact, the price instability would be eliminated.

If such a program were put in place, the transformation of the U.S. industry would begin almost immediately. The U.S. automakers would be able to predict what products would be in demand when the gasoline prices reached a certain target level and they could start adjusting their automobile production accordingly. Americans could plan their future automobile purchases similarly.

There are many modifications possible. For example, a modification could be made when taking into consideration the gasoline behavior in Europe. European nations now pay around $7.00 per gallon of gasoline, yet they continue to make the most of high efficiency gasoline automobiles. There is no evidence that Europeans are turning away from gasoline automobiles in large numbers, demanding electric vehicles or some other gasoline alternative; instead, Europeans are beginning to pay attention to the price of gas and are slowly beginning to explore other alternatives. Nevertheless, for now Europeans are comfortable paying these high prices and no one knows how high the price will need to go before the interest to look elsewhere is sought on a large scale.

It may be concluded, given the European example, that Americans may behave in a similar manner. It may well be that in order to get any kind of public attention, the U.S. would need to speed up the initial price increase and adjust the price of gasoline at a somewhat faster rate. Perhaps a one-dollar increase per year for the first 5 to 7 years, and 10 percent thereafter, or bringing the price of gasoline to an even par with Europe all at once, would be more appropriate. Again, anyone can fill in the blanks, as to what the actual numbers should be. Should the rate be 12 or 15 percent per year or should the program span 20 or 30 years. Filling in the blanks is simple; getting the plan started is the hard part.

Once started, it becomes an easy task to predict what effect the program will have at certain price target milestones. Assuming the original example and assuming Americans would opt for this kind of solution, then:

$5 gasoline at the end of 7 years: this price target would bring oil to an even par with most alternative fuels. Advance oil recoveries would become profitable and the oil companies could begin increasing the oil production, which is now on a downward trend. Hopefully, getting back to production of 10 million barrels of oil per day, a level at which U.S. production peaked, could become the new goal of the oil industry in this country. At a $5 level, production of biofuels and ethanol becomes a profitable business and no longer requires any farm subsidies. Discussion of high fuel efficiency automobiles would make Americans start noticing high fuel efficiency hybrid automobiles. Although it is unlikely that high fuel efficiency hybrids, especially the compact models, would increase dramatically in popularity at the $5 gasoline price levels, the mere fact of it being discussed would make Americans more familiar with the hybrid concept. Americans would begin to accept the hybrid concept as a reasonable alternative, rather than just laughable cheap foreign imports.

$10 gasoline at the end of 14 years: at a $10 level, farming would be on the rise. Planting corn and soybeans to produce ethanol would become a very profitable business. Of course, ethanol production could never increase in sufficient enough amounts to replace oil completely; however, if the portion contributed to the overall oil production amounted to only 10 or 20 percent, it would contribute to a substantial reduction of foreign imported oil. Americans would begin to demand alternative fuel sources for their automobiles to reduce the cost of driving. High fuel efficiency hybrids, including the plug-in electric hybrid automobiles, would dramatically increase in popularity and would be sought by many Americans to replace the standard gas-guzzlers. Plug-in electric-gasoline engines would become commonplace. With expensive gasoline, Americans would be much more willing to charge their car batteries, even overnight if necessary, by using the existing electric grid. They would begin to use the inexpensive electric power in their automobiles for everyday driving.

With the plug-in hybrid technology on the rise, consumers would demand longer-lasting batteries. Research in this technology would likely be accelerated to shorten the charging time significantly, where minutes would replace hours. Americans would demand technologies to replace gasoline, and if the U.S. automakers refused to provide the U.S. consumers with what they wanted, then the foreign competition would certainly fill in the gap.

$20 gasoline in 21 years: by the time the price reached $20, the transformation would most likely be complete and the oil dependency would be defeated. Most Americans would be driving either high fuel efficiency hybrids, plug-in electric hybrid automobiles, electric-only vehicles or maybe even hydrogen-powered automobiles, while utilizing biofuels, biodiesel and ethanol in place of gasoline. Somewhere before the price reached $20, the U.S. oil consumption would likely drop to the point of being self-sufficient, which would mean that the United States consumes no more than 10 million barrels of oil per day. Once self-sufficient, Americans could continue using

When All Else Fails (The Europe Way)

up these domestically produced quantities for many generations.

The policymakers who would provide the price increase structure would certainly have numerous options to design a system that would provide optimal results. They may conclude that the 10 percent increase may be too much or too little. All industries have to be considered and proper time for the plan has to be allocated. If too much time is allowed then all the work attempted may be for nothing, in which the realization of shortages may precede any meaningful solution. On the other hand, if not enough time is allowed, then the U.S. industry may not be able to absorb all these changes painlessly and the unwelcome consequence of crippling the economy and a prolonged economic slump may be realized. A 7 percent yearly increase would double the price of gasoline every ten years and the $20 gasoline would not be realized for thirty years. A 15 percent increase would double the price every five years and $20 gasoline would be a harsh reality in only fifteen years.

To illustrate what happens if the program moves ahead too slowly, all you need to examine is the past, the last thirty years to be specific. In early 1975 gasoline was around 35 cents per gallon. If a 7 percent yearly increase were applied then, the gasoline would double 3 times in 30 years. A price of 35 cents in 1975 would reach 70 cents in 1985, $1.40 in 1995 and $2.80 in 2005. That is pretty much what Americans pay now; therefore a lot of effort for nothing. A 10 percent increase would reflect the price Europeans pay today. A 10 percent increase would double the price of gasoline 4.1 times in 30 years, which would bring the price of gasoline to around $6, comparable with the $6 to $7 Europeans pay today. This is little better, but still a sad consolation prize considering the effort that has been put in. A 12 percent increase would double the price every six years, or five times in thirty years, to a price of $11.20 per gallon; a 15 percent increase would double the price of gas every 4.8 years, which would double the price of gasoline 6.25 times in thirty years to $28 per gallon in 2005. So to take a lesson from the past, a nice round number for the future may just be in the 10 to 12 percent range.

Side by Side with Europe

Should the U.S. government take the European path instead of one of the other conservation alternatives, then the price of gasoline should increase to match Europe's. That is to increase the price of gasoline to $6.00 or $7.00 immediately, all at once, rather then with the 50 cent tax that exists today. Then continue side by side with Europe, kind of as their partner, to increase the price of gasoline similar to the way they do. Or adopt the 10 to 12 percent increase per year. Don't be flabbergasted because the odds are not on the U.S. side.

"If the U.S. does not raise the price of gasoline first, and take advantage of the collected funds, then OPEC will raise the price of oil for the U.S., collect the funds and laugh all the way to the bank."

These are the facts. If the U.S. government does not increase the price of gasoline first, OPEC will do it in the next few years. Considering how the current world events are developing, the U.S. may see $10 gasoline by the end of

When All Else Fails (The Europe Way)

this decade. If the U.S. does nothing, Americans may indeed see OPEC countries laughing all the way to their banks. So what is it going to be?

There is a choice to make. One choice is to do nothing. Let the chips fall where they may and allow the market forces to produce an adequate solution. The second choice is to establish a program, which will provide the U.S. with a solution now. Both scenarios will lead the U.S. toward much higher oil prices. The first scenario will suck out all the money from the Americans and send every single dollar (trillions of dollars) overseas to the OPEC countries in the Middle East, Venezuela, Russia, and Africa; fattening their bank accounts and leaving Americans with only gas fumes after all of the oil has been burned up, while creating uncertainty, shortages, confusion and chaos along the way. The second scenario is, while the price of gasoline increases similarly through artificial means, the dollars will remain in the United States for Americans to spend as they want.

This should be the primary driving force for the U.S. government; this should be the government's incentive to come up with a solution. If the price of oil is going to increase dramatically in the near future no matter what, and will cause Americans to spend a much larger proportion of their disposable income on gasoline, no matter what, then why not increase the price of oil artificially and subsidize alternative fuels abundant to the U.S., to help with the transformation from oil economy to alternative fuel economy. There is no shortage of promising ideas, just a shortage of resolve to select one idea and proceed.

What should the U.S. government do with the ever-increasing revenue that is going to be created? The European solution collects large sums of money. For every dollar increase the U.S. would collect over $300 billion in revenue, and with gasoline at $7.5 per gallon, where $5 is in the form of tax, $1.5 trillion in revenue would be collected in one year and that would go a long way to eliminating the U.S. deficit and having plenty of funds to spend on other things. In fact, if these funds were applied to paying off the U.S. debt, it would take only a few years to eliminate the debt completely. What would the U.S. government spend these funds on? I suppose that is for the lawmakers that had the guts to implement the new policies to decide.

Gasoline was used in this example. Unfortunately, the solution may not be that simple. While the system of using gasoline provides for simple accounting, it leaves the U.S. oil industry without any direct price benefits. Without additional revenue from the existing oil production, the oil industry may lack funds to use in more expensive research, such as oil shale and the enhanced oil recovery of existing wells. Some form of subsidy would need to be provided to the U.S. oil companies for the portion of crude they produce domestically. This would put the U.S. producers on an equal footing with all other domestically produced forms of energy.

None of the programs discussed are designed to make life miserable for Americans, rather they are trying to prevent an economic disaster from occurring some time in the future, which may be several decades away or could pop up in a year or two. And although the electric solution, or one of the conservation ways would probably be the preferred way of the U.S., the Europe solution (the bombshell approach), no matter how distressing, no matter how harsh and merciless, is a really well-thought-out solution that would certainly accomplish oil independence.

9.
Fun with Numbers

Alternative Solutions

Oil alternatives have been the subject of scrutiny to many. Individuals, including scientists, provide the American public with information of energy alternatives in absolutes. Absolute analysis scrutinizes individual energy alternatives on a basis of whether they have the potential to replace oil completely. Some of these people even go so far as to base their analysis on whether the particular alternative can replace all of the fossil fuels currently in use in the United States. These arguments usually point to the impracticality of one energy alternative over another by providing the figures needed to replace all of the energy in the U.S. or the world. In this type of absolute analysis, these individuals usually promote one source of energy while trying to directly or indirectly discredit the other types of energy. Absolute analysis is almost always negative, yet it is used over and over in many articles.

Wind and solar power is almost always found inadequate due to the vast area needed for wind and solar farms. Both of these sources are regarded as inefficient due to their intermittency. The argument goes that wind generators cannot run without wind and because they provide only 30 to 35 percent efficiency and need three times as much wind power capacity as nuclear, coal or hydropower, they are inadequate for large scale applications. Solar power is regarded as having even smaller efficiency ratios, thus the same conclusions. The apathy to solar panels stems from the fact that they cannot generate power without the sun, or on cloudy days, and that they are only 10 percent efficient in collecting the sunshine energy, therefore the solar panels are written off as not having any potential at all.

Nuclear power is considered very unsafe. Activists and politicians regard building 1,000 nuclear power plants that could replace oil with electric power as irresponsible and dangerous. Coal power is considered too dirty and the environmentalists would have a field day pointing out the huge amounts of CO_2 emissions, if someone even remotely suggested building 1,000 coal-fired power plants to replace oil. And finally, trying to replace oil with biofuels provides yet another conflict, that of available farmland. It is also argued that it could increase food prices, if enough farmers decided to grow crops for ethanol production rather than food.

And so, in the spirit of curiosity, what follows will provide a few simple calculations in which individual alternatives will be considered to replace oil. Each one will be considered to replace all of today's imported oil, the 14 million barrels per day. Later, all of the alternative fuel options will be combined. Rather than deal in absolutes, one fuel being capable of doing it all, attention will be given to how a small increase in each individual alternative could result in an effortless journey to oil independence.

Electric Power Plant Production

1 barrel of oil = 1,700 KWh or 5.8 million Btu

14 million barrels of oil per day equals 5,110 million barrels of oil per year.

5,100 million barrels of oil x 1,700 KWh = 8,687,000 million KWh. (8,687 billion kilowatt-hours)

Thus, to replace the 5,100 million barrels of imported oil each year, 8,687,000 million KWh of electric power would be required.

A new modern nuclear or coal-fired power plant has an average capacity of 1,000 MW of power, which is equal to 8,760 million KWh per year.

Doing the math, 991 new power plants would be needed to replace all of the imported oil.

Assuming power plants are 80 percent efficient, 1,238 power plants would be needed. A nice round number may be 1,250 new power plants.

According to the Department of Energy, U.S. electric consumption today is 3,970,555 million KWh per year, with the generating capacity of 875,870 megawatts or 7,672,000 million KWh per year.

Wind Energy

One MW wind turbine operating at 33 percent capacity generates 2.89 million KWh of power per year.

8,687,000 million KWh would be needed to replace imported oil.

Doing the math, 3,006,000 wind turbines are needed to replace all of the imported oil.

Provided 3-megawatt turbines were used instead, then 1 million wind turbines would be needed to replace all of the imported oil.

Solar Power

Solar panels, which are only 15 percent efficient today, produce .45 to 1.35 KWh per square meter per day.

With 2,000 square feet (185.8 square meters) of roof surface on an average home, this means that each house can provide 30,529 to 92,102 KWh per year. To simplify, the average value is 61,315 KWh per year.

In the U.S., with approximately 70 million homes, the total capacity of rooftops is 4,294,500 million KWh per year.

With 8,687,000 million KWh needed to replace all of the imported oil, the solar panel efficiency would need to double to 30 percent, to use rooftops only for solar power generation.

In a desert area with sun exposure of 1.35 KWh per square meter per day, the area needed would be: 1.35 x 1,000,000 (meters square per kilometer square) x .3861 (miles square per kilometer square) x 365 days = 190.25 million KWh per square mile.

To produce 8,687,000 million KWh; 45,660 square miles would be needed, which would cover an area of 213.6 miles long by 213.6 miles wide. That may not be that bad if one considers the usefulness of a desert property.

Biofuels or Ethanol

Only slightly more than 1 percent of the fuel today is derived from ethanol. When one considers corn as the primary crop for ethanol production, farmers would have to produce nearly 50 times the amount of corn they produce today to replace the 14 million barrels of oil per day.

2.7 gallons of ethanol can be produced from one bushel of corn, and one acre yields approximately 140 bushels of corn.

Thus each square mile provides 2.7 x 140 x 640 = 241,995 gallons of ethanol

241,995 gallons of ethanol / 42 gallons per barrel = 5762 barrels of ethanol per square mile, or .00576 million barrels per square mile.

To replace 5,110 million barrels per year:

5,110 MB per year / .00576 MB per year = 887,153 square miles.

This means that an area 942 miles wide by 942 miles long would need to be planted with corn to replace all of the imported oil.

Alternatives Combined

This section illustrates how it all could come together. No longer will a "winner takes all" discussion be employed. Combining all of the available U.S. resources together is when the enormous U.S. potential can finally be realized. When this country determines that importing oil from abroad creates problems and forces sacrifices that Americans are unwilling to make, the U.S. government can simply put a stop to it. The U.S. does not need foreign oil. It has an abundance of its own resources to replace all of the imports now and to produce sufficient energy for the American people for thousands of years to come. First, the U.S. needs to increase new oil production by establishing a floor price of oil. Second, just as important to any solution, the U.S. has to commit to energy conservation. Let's discuss what U.S. has at its disposal.

Increase the oil production

Establishing a floor price for gasoline that the U.S. oil companies must charge is probably the most important part for an overall plan to increase oil production, yet it may turn out to be the most controversial. Why would someone want to guarantee a higher price of oil to the oil companies when they are already making record profits? A $100 barrel of oil translates to $3.50 to $4.50 per gallon of gasoline. A government mandate establishing a minimum price of $4 per gallon of gasoline, for example, would provide an incentive for new oil exploration, revisit the existing abandoned oil fields and provide an incentive to engage in the production of oil from the abundant supply of American oil shale. With a guaranteed price of oil, the U.S. oil producers are guaranteed profits. It would not be too long before either the oil companies or a new business enterprise discovered the oil shale potential.

It is true that oil shale, for example, could be profitable at crude oil prices of $50. But it is not just the cost of production. A new industry requires a completely new infrastructure. This type of large-scale operation would involve building new refineries, roads and railroads. It would require plentiful water access for the production. It would require building new housing, schools and hospitals, and millions of dollars spent on large monstrous machines. Huge amounts of money would have to be invested before one gallon of gasoline was produced. Most of the other alternatives require similar investments. Unfortunately, no one wants to invest large sums of money to see the price of oil drop and the new operation become unprofitable. Thus the floor price of a barrel of crude oil would become a great incentive for new production.

It is conceivable to think that by combining the renewed interest in oil exploration with the existing production, the U.S. could once again produce 10 million barrels of oil per day; the amount the U.S. used to produce before the cheap Middle East oil flooded the U.S. markets. With the current oil production of 7 million barrels of oil per day, that adds 3 million barrels per day, which is certainly possible once the profit margins increase. Using Canada as an example to what may actually happen; Canada estimates to increase its oil production from their tar sands. Their current production of less than one million barrels of oil per day is projected to rise to 3 million barrels per day within the next ten years. Therefore, the U.S. production estimates of 3 million barrels per day from the U.S. oil shale, especially when combined with the increased interest in the existing abandoned oil fields, are not exaggerated. Certainly no more than ten years would be needed to insure a new booming oil industry.

Another significant source of oil production increase is biofuels, specifically ethanol. Being liquid, it is the only renewable resource that can physically replace gasoline itself. Unfortunately, the production costs continue to be too high to compete with current gasoline prices, and therefore ethanol has not been seriously pursued. In 2005 ethanol provided around 1 percent of the total U.S. auto fuel. With a minimum price of oil at $4.00 per gallon, the U.S. ethanol production could be expected to skyrocket. In any case, ethanol production has the potential to easily supply up to 4 million barrels of oil per day within ten years, without too much of an effort. Additionally, if the lawmakers established a gasoline additive requirement, say 20 percent, and adopted a target year when the gasoline additive became mandatory, the program could be jumpstarted. The farmers could expect generous financial investments from the oil industry, which due to the additive requirement would have direct interest in ethanol production, and the oil industry could spend some of the enormous profits they currently enjoy.

None of the oil has been eliminated so far; only the ratio of imported oil has changed. Of the total 21 million barrels of oil per day, 14 millions were imported and 7 million were produced domestically. Provided the two changes discussed above came to pass within the next ten years, with the 3 million barrels from increased oil production and the 4 million barrels from the increased ethanol production, the ratio would reverse to 14 million barrels of oil produced domestically and 7 million barrels of oil coming from foreign imports; now only 7 million barrels per day would need to be addressed and eliminated.

Not to overdo the savings, the U.S. should not forget their North American partners, Mexico and Canada, the "North American OPEC," if you please, who supply the U.S. with 4 million barrels of imported oil per day and with Canada expecting to increase its production from their tar sands by another two million barrels per day, the U.S. could expect to receive some 6 million barrels of oil per day within the next ten years from these neighbors. Even though these are "imports," these countries are neighbors of the United States so their supply should be considered safe. That leaves a mere one million barrels per day to go.

Conservation Control Commitment

The many conservation options that have been discussed would provide a way to reduce a very significant portion of imported oil. It just so happens that the 14 million barrels per day which the U.S. imports now is the amount of oil used in the transportation sector, of which close to 9 million barrels per day is used by the light-vehicle transportation sector. Just by doubling the light-vehicle fuel efficiency from 20 mpg to 40 mpg would cut the consumption in half and would save 4.5 million barrels per day. By further increasing the fuel efficiency to 60 mpg, we could cut the consumption by two-thirds, conserving 6 million barrels per day. That is a good start using conservation, greatly surpassing the goal needed. What's more, this could be easily done within the next fifteen years or less.

And should the price of oil increase, be it naturally or through artificial means as discussed previously in the Europe solution, the U.S. could expect a reduction of oil use in all of the other industries. The rest of the transportation industry such as commercial light trucks, freight trucks, rail, shipping and bus transportation would also reduce its oil use, in addition to the potential savings of plug-in hybrid technology. Additionally, increased oil prices would certainly reduce oil use in industrial, commercial and home heating, and encourage substituting the oil with other types of energy, most likely electric. The increased price of oil would certainly be responsible for saving a portion of the 10 million barrels per day now used in these sectors. How much of a reduction in use could be realistically expected? Not to keep guessing, but a 20 percent reduction would save 2 million barrels per day, and with every additional 10 percent savings there would be another one million barrels of oil saved. With the commercial industry continuously striving to improve efficiency, it would not take these companies long to replace the expensive oil with significantly cheaper alternatives.

To sum up, there is more than enough oil available in the U.S. to easily part with OPEC. All that is needed is a little resolve and determination on the part of the U.S.

government to produce a simple but serious plan. Consequently, there is no need to have the U.S. Army scattered throughout the world. Now that the potential oil supply has been rediscovered, let's see if electric production can follow suit.

Electric Production

The electric power use in the U.S. is expected to continue to increase regardless of whether or not electric is considered to replace auto fuel. Electric demand is expected to increase by 45 percent by 2030 and coal is expected to take on the bulk of the increase. By now it should be evident that electric power will not need to replace all of the imported oil, as previously thought. The electric requirements will be much smaller. Some electric would be needed in the transportation industry; however, the bulk of the new capacity will be needed in other industries for use in heating and cooling. How much can be reasonably expected from the various electric alternatives? Let's examine how much effort is needed if each of the electric producers was expected to replace 3 million barrels of oil per day within the next ten years.

Electric Surplus

The U.S. currently consumes a little less than 4,000,000 million KWh per year and has a little less than 8,000,000 million KWh of total capacity. As discussed earlier in the section on absolutes, the U.S. has a large surplus capacity. If consumers found the way to use this extra capacity through some form of incentives given by the electric utility companies during the off-peak hours, this extra capacity could be utilized. Replacing 3 million barrels of oil would require 1,861,500 million KWh, which is approximately half of the existing excess capacity.

New Power Plants

Suppose that the U.S. government would be uneasy about jeopardizing the excess capacity in any way and would opt to increase the electric production instead. Assuming an 80 percent efficiency rate, then 89 power plants would be needed to replace 1 million barrels of oil per day. Thus, building 267 new power plants divided between nuclear, coal and hydroelectric, in any combination, would replace 3 million barrels of oil per day. It should be noted here that the U.S. has enough coal reserves to generate enough electric for the next couple hundred years. Also, enough potential resources of uranium exist to generate the needed electricity for millions of years. Lastly, hydropower is naturally renewable and can last indefinitely.

Wind

In 2005, California had 13,000 wind turbines producing electricity. Around 200,000 turbines, 3-megawatts each, would be needed to replace 3 million barrels of oil per day. The great asset of wind turbines is that they can be put into production one at the time, 3 megawatts at a time, rather than the 1,000 MW of total capacity for a new modern power plant. This ability makes the 200,000 much more manageable. Supposing the production capacity to replace 3 million barrels of oil would be needed in ten years, then electric utilities will need to put into production just a fraction of the 200,000 turbines needed. This amounts to 20,000 per year, divided between the thousands of power plants in existence in the U.S. today. Suppose only the one thousand largest power plants were required to add this capacity. That would only involve 20 wind turbines per plant per year to meet the desired goal.

Solar

As inefficient as solar power is today, with only a 10 to 15 percent efficiency of collecting sunlight, an area that is 100 miles long by 100 miles wide that is tucked away somewhere deep in a desert and covered with solar panels could replace the 3 million barrels of oil. Utilizing rooftops by having a mandatory requirement of solar panel installation on all new construction, and providing incentives to existing homes, can easily generate enough new solar electric power to replace an additional 3 million barrels of oil per day within the next ten to fifteen years. And as solar panel efficiency improves, the solar electric power generation could actually be much more successful than previously thought. Again, solar power, like wind power, can be added incrementally as needed, therefore able to utilize the most advanced solar power technology as soon it becomes available. Who ever said solar had no future?

By hypothetically combining the electric capacities that can be added to the current production within the next ten to fifteen years with relatively minimal effort, enough electric potential has been found to replace 15 million barrels of oil, which far exceeds the projected demand increase.

More on Conservation and Efficiency

Recall the soccer mom example, in which the fuel needed to take a 120 pound person to a destination is 70 times higher just because the 8,500-pound SUV had to be taken along? There are many other examples in which the waste is just as great. The government could get involved in conservation on many other fronts. Conservation and efficiency does not only pertain to oil and the automobile; conservation could be applied in many other every day applications from light bulbs, refrigerators, furnaces and water heaters to better building codes providing better insulation, window replacements in older homes, assistance with installation of solar panels and many others. With the absence of pure partisan politics, the government could take on a leadership role in many of these areas. There are many environmentalist organizations that are constantly educating American public and the U.S. businesses to conserve. Unfortunately, all this effort has a minimal effect on the overall conservation. The U.S. government, on the other hand, just by passing few simple conservation laws, would have an exponentially greater success.

Energy efficient light bulbs have been on the market for many years. Many of these are four times as efficient as their incandescent counterparts yet the energy wasting light bulbs are still being sold throughout the country. Okay, maybe light bulbs will not make much of a dent in total energy saved, but when combined with all of the other energy savings the quantities will start adding up. Most U.S. homes have refrigerators, water heaters, furnaces, TVs, air-conditionings and many other appliances. Suppose that the Department of Energy mandated that manufacturers have to increase the energy efficiency of all home appliances by 7 percent per year; in ten years the energy use of these appliances would be cut in half and in 20 years by three-fourths; that is conservation worth noticing.

"Two for the price of one, or buy one get one free." Everyone is familiar with this phrase. It needs no explanation. It is an accepted everyday term in every store and every restaurant. Buy one pizza, get one free. Buy one pair of shoes, get the second pair free. The same applies to conservation of energy and improved efficiency of everyday products. If all of the home appliances, equipment, and household tools were two times as efficient, then consumers would end up using only half of the energy; thus, able to use twice as many gadgets for the same price of energy.

Many Americans own old homes and are familiar with single-pane windows. To conserve, many of these homeowners have gone through the trouble to replace these windows on their own in order to save energy. Yet the building codes for new homes still allow for single-pane window installation. As of 2006, Florida builders still regularly install single-pane windows on brand new homes selling for upward of $400,000. If the Department of Energy was the least bit interested in energy savings, a mandate demanding triple-pane energy efficient windows could be implemented overnight. Additionally, the government could, rather than helping to pay the energy bills of many of the poor homeowners, provide help to insulate the old homes and replace the old windows, thus helping these homeowners by lowering their energy bills.

Initially, by helping to replace the old windows with more efficient windows and by helping with insulating the old homes, the government's upfront cost would be higher than if they just helped to pay the high energy bills; but the cost would be incurred only once. Once the home is insulated and the new windows are in place, the government's job is finished. The homeowner enjoys affordable energy bills and the government can move on. With the energy assistance "invention," on the other hand, the government has gotten involved in paying huge amounts of money year after year; a never ending cycle, and not an ounce of energy is saved in the meantime. It is like the old saying "Give a man a fish and you feed him for a day; teach him to fish and you feed him for life." Thus the politically motivated energy assistance invention involves always giving.

Conservation and efficiency usually have a negative connotation and usually imply giving up something or lowering the standard of living. Here, however, what has been shown is that conservation and efficiency mean something completely different. The ability to use two gadgets for the price of one; that stands for a better standard of living, not worse. Conservation and efficiency does not require a lot of money, it just requires passing a few simple common sense laws to stop the energy waste.

Conclusion

*N*ow that the end of the oil journey is approaching, how can it all be put together? Everyone is in agreement that as the public dismay grows when the price of oil increases, something needs to be done. What is currently lacking is a plan of how to begin. What follows is an example of a simple plan, in few simple steps; and again, you, the reader, may alter it any way you feel, if you come up with something better.

First, the U.S. has dug itself into a deep hole, and so as the saying goes, "if you dig yourself into a hole, the first thing to do is to stop digging." Dealing with the U.S. petroleum imports, the first thing to do is to end the increasing petroleum consumption; that means no additional oil imports. The 14 million barrels of oil the U.S. currently imports has to become the absolute maximum of U.S. oil imports. In order to accomplish that, the U.S. needs to artificially, by using a gasoline tax, raise

Conclusion

the price of oil, where this country can at last begin to live within its means. The tax may need to be significant, like $2 or $3, and maybe even higher to establish equilibrium between supply and consumption. Furthermore, the tax may need to be increased anytime that the demand begins to rise again; which, by the way, could very possibly happen once Americans become accustomed to the new oil price. To reiterate the statement used previously, "If the U.S. does not raise the price of gasoline first, and take advantage of collected funds, then OPEC will raise the price of oil for the U.S., collect the funds and laugh all the way to the bank." It is America's choice.

Once the bleeding has stopped and the U.S. begins to live within its own means, a bill should be passed to begin producing electric plug-in hybrid automobiles, and any new production of gasoline-only automobiles should be outlawed. This action would, within the next fifteen years or less, transform most of the U.S. auto fleet to where Americans could take advantage of domestically produced, inexpensive electric power. The use of this power would bring driving costs for most Americans down to less than one dollar per gallon of gasoline equivalent; a price range at which all Americans, including all of the U.S. business community, would flourish.

When the transition of the U.S. auto fleet to electric hybrid plug-in vehicles is well on the way, action should be taken to begin reduction of oil imports. This can be accomplished through the various conservation efforts introduced in the conservation chapter. Increasing CAFÉ standards by 10 percent, or 3 miles per year for example, would provide a fleet of automobiles not only capable of using electric fuel as a primary fuel source, but when gasoline is needed, the fuel efficiency is high, 80 mpg or even more.

Finally, adding the Europe solution would bring about the completion of the process. Starting in ten years or so, when most Americans have had the chance to switch to dual fuel vehicles and have the ability to use alternative inexpensive fuel, the U.S. should begin to increase the petroleum price by 15 percent annually, until equilibrium of production and consumption is reached. This would virtually assure complete independence at the end of twenty years.

Therefore, the implementation of four simple steps could transform America from an oil addict to an America that depends on no one; an America that does not owe anything to anyone; an America that is self-sufficient in all of its energy needs; an America where everyone enjoys inexpensive abundant fuel.

Is this a fantasy or could this scenario become a reality? Where is the U.S. going to be in the next twenty years? What are the chances that the U.S. Congress and the president will cooperate to implement some kind of solution to achieve oil independence in the foreseeable future? It remains to be seen. Brazil is proving it can be done. Iceland is showing a hopeful promise that it can be done. The European Union is doing it their way. In other words, all of these countries are taking advantage of their natural resources and sending a message to OPEC that they will no longer continue to depend on their precious oil. So where does the U.S. stand? When is the U.S. going to decide to do it their way?

For now, everyone is out there doing his or her own thing. Oil companies will not initiate anything because they are driven by profits. Automakers will not initiate anything because they are driven by profit. The politicians are out there driven by reelections.

The oil arguments in the U.S. government are in their infancy. Actually, they did not even have their debut. For now the squabble just evolves around the argument of big oil business vs. small consumers. Eventually the oil crisis will arrive and ultimately the discussion of oil prices will become serious, and once the oil reaches a high enough price, the dispute will take on government partisanship. Some in the government will argue on behalf of the oil companies, some will argue on behalf of Detroit, some may even argue on behalf of OPEC itself, and a minority in the government will argue on behalf of the U.S. consumers. Passing a few simple laws to eliminate oil independence will become a difficult and cumbersome process. Let's examine how laws in the U.S. get passed and see if there may be something that will make this a more simple and uncomplicated process.

Many laws do get passed. It is actually a straightforward but somewhat lengthy process that involves cooperation of all branches of government. First, a member of Congress introduces the bill, which can be done at any time the House is in session. After the bill is read on the house floor, the bill is referred to one of the many committees. The appropriate committee reviews and debates it; it makes changes to the bill and then votes on it, either killing it or sending it back to the house floor. Sometimes it is sent to a subcommittee for additional study. A subcommittee also has the power to either kill the bill or revise it and send it back.

Assuming the bill gets through the committee, it is sent back to the house floor, where house members debate the bill and offer amendments and vote on the final draft. If the majority passes the bill, it is referred to the senate to undergo similar scrutiny. Again the bill is sent to a committee where it can be amended, killed or sent back to the senate floor and voted on. The senate must also pass the bill in order for the bill to become a law. If the language of the bill is changed, then it must be sent to the conference committee, which is made up of members of both the house and the senate. All the differences must be settled before the bill is sent to the president.

The president may sign the bill and make it a law, or he may decide that the bill is bad or unnecessary and issue a statement of objections called a veto. If the bill gets vetoed by the president it goes back to the house, where it will be read and a decision will be made if it should die, in which it does not become a law, or if a vote should be taken to override the veto. A two-thirds vote is needed in both the house and the senate to override the veto. A successful vote will allow for the bill to become a law. In the case of an unsuccessful bid to override the veto, the bill is killed.

From the description above one can see that the bill has to go through many stages and can be killed any time along the way. Yet many laws do get passed. The bill may enter the process as a partisan issue and through cooperation and a little give and take it may emerge at the end of the process as a piece of legislation with bipartisan support.

Currently both houses are divided equally between republicans and democrats, so the president is usually consulted to avoid a veto. One interesting point to add is that if both parties in the house and senate decided to cooperate, then the president's position becomes irrelevant. In fact, because the law originates with the Congress that enforces the passage of the law with two-thirds vote, the president is not needed at all. As mentioned previously "what an awesome power!" Although neither party can do it on their own, law making is simple and Americans should insist that all issues get resolved in a bipartisan and timely manner at the end. There is no reason why Americans should tolerate Congress keeping so many issues unresolved for such a long time. It does not serve the public, just the politicians.

Compromise is the name of the game. Through a compromise every issue, even the most difficult ones, can get passed in a matter of months. Once Americans figure this out, they will be able to kick out all of the unwilling-to-compromise politicians of both parties, the hard-liner obstructionists that do not care about anything or anybody.

One reason the issues such as immigration, abortion and many other religious and non-religious issues remain unresolved is because the lawmakers get a lot of political mileage out of them. It is the bread-and-butter of the political process. If these concerns got resolved, the issues would become politically dead and the politicians would have to focus on other topics to base their campaigns on. And so the charade continues and politicians get to run on these issues over and over again, promising to resolve the problems but actually never having any intention to do so. It sure seems that due to partisanship there is a real roadblock to pass even the simplest of the laws. Would these roadblocks interfere with passing laws to free America of imported oil?

On the path to becoming oil independent, the U.S. should also plan for the future. As mentioned previously, hydrogen could easily become "the new kid on the block" in some 20 or 30 years. Research should be expanded so hydrogen becomes a reality quickly. Hydrogen could take its place alongside the other primary fuels. The automobiles of the future would not only include electric-gasoline hybrids and the plug-in hybrid electric vehicles but they would include hydrogen-gasoline, hydrogen-ethanol and hydrogen-electric plug-in vehicles as well.

A hydrogen-gasoline combination vehicle could easily be incorporated into the U.S. fleet of automobiles. This could be done similarly to the electric-gasoline automobile described in chapter 6. Rather than storing enough hydrogen to cover several hundred miles in a hydrogen-only automobile, the hydrogen-gasoline automobile could easily store enough hydrogen to cover 100 miles or so, and gasoline or ethanol could take over from there. Additionally, if hydrogen was cheaper than gasoline, hydrogen could be refueled during long trips unlike the electric power in the electric-gasoline counterpart.

The development of hydrogen would certainly create an additional safety net for the availability of fuel for automobiles on the U.S. Highways. Hydrogen can virtually be made in unlimited quantities without adding to global pollution; however, hydrogen needs electric power to be produced. Assuming that we exclude the production of hydrogen from fossil fuels, then in addition to the off-peak electric power supply, supplied by the nuclear and hydroelectric power plants, there is the solar and wind power. Even the biggest skeptics, who criticize solar and wind power intermittency problems, would have to be satisfied if the solar and wind energy was used to produce hydrogen by the process of electrolysis. Hydrogen would be made only during nice sunny days when solar power is abundant, or during the nice windy days when windmills would be utilized at their optimum levels. If more hydrogen was needed, simply adding a few more solar panels or a few more wind turbines would resolve the dilemma, thus an unlimited and pollution-free source of fuel that could be used forever.

Finally, when discussing the future, one must not exclude the previously discussed biodiesel obtained from algae. Just as hydrogen, this fuel source could be sufficiently advanced, to produce biodiesel in large quantities in some 20 to 40 years. Algae can produce up to 10,000 gallons of fuel per acre. Replacing 3 million barrels of oil per day (1,095 million barrels per year) would require an area of 7,186 square miles, or an area of 85 miles wide by 85 miles long. Keep in mind that an area of 190,100 square miles, or 436 miles by 436 miles, is needed to produce the same amount of fuel from corn. If the same 190,100 square miles would be devoted to the production of algae, then that would produce enough biodiesel to replace the 80 million barrels of oil per day, (29,200 million barrels per year), which just so happens is the current total world oil consumption. This kind of undertaking, though immense in scale, has the ability to do away with crude oil as we know it. That just may be something to invest in for America's future.

In the future, gas stations will likely take on a new function too. Today, customers get to select from a variety of products, all of which are made from crude oil. These are regular, premium and super unleaded gasoline and diesel fuel. Future gas station will likely have a variety of choices too, but they will be different. They will likely be replaced by the following: In addition to gasoline made from crude oil, there will be a variety of other man-made fuels like hydrogen, ethanol and biodiesel. In addition to these fuels, the gas stations of the future will likely have electric battery charging stations, where customers will be able to charge their batteries in a matter of minutes, rather than hours, as will continue to be the case of home charging systems. Thus it will make absolutely no difference what fuel the automobiles of the future use. Whatever the fuel is, it will be available in abundant quantities in the U.S. gas stations. It should be noted that just as the diesel fuel of today, which is only available in a few selected gas stations, the new fuels of the future could be available similarly. There certainly would be no need for each gas station to stock every available fuel.

The future of home heating is likely to change drastically too. As the new solar panels and wind turbines become more sophisticated, they will be installed on the roof of every home, providing electricity to heat and run the homes. Due to high efficiencies that will be available in the future, most of these systems will be able to produce more electric power than the home can use. Homeowners will be able to use this excess power for transportation, to charge the batteries or perhaps even to make their own hydrogen, and the rest of the power may be sold to the electric utilities. The electric utilities will be able to resell this power at a profit to other individuals and industries that do not have self-sufficient capabilities.

Conclusion

In the not-too- distant future, dirty coal can begin to be phased out too. The old outdated coal-electric power plants can begin to be replaced by new modernized coal power plants with scrubbers to clean the smoke, thus eliminating the other major source of atmospheric pollution contributing to global warming. As new technologies are being put into place, coal could be phased out altogether, being replaced by the pollution-free sources, ending the plight of the environmentalists.

Global warming has finally been recognized as being created by burning hydrocarbons and is today gaining the attention of the environmentalists and the government officials. Many plans introduced by these groups strive to reduce the pollution being released to the atmosphere by just a few percent, finding even that almost impossible to accomplish. By reducing the consumption of crude oil from the existing 21 million barrels per day to 10 million barrels per day would reduce the pollution released to the atmosphere by a substantially larger margin, and if combined with the building of new modernized coal power plants only, the rate of the global warming effect could be reduced by at least half in the next 20 years. Of course the long-term goal should continue to be to reduce the atmospheric pollution altogether.

To sum up, what has been recognized is that the future of energy is marvelously bright. What has been revealed in the few pages here is that the society of the United States and the whole world is on the verge of a magnificent transformation. Nations are on the threshold of transforming from the Stone Age, in which they burned decomposed prehistoric animals and plant remains, "the fossil fuels," to the modern era, in which they begin to harness the limitless energy of the sun itself and in which they can recover the boundless energy from the wind blowing around them.

Humankind is on the verge of exploiting the virtually unlimited geothermal energy of the interior of the planet and they are on the verge of splitting atoms to create practically infinite amounts of energy, using nuclear power. Humankind is on the verge of splitting water by using only sunlight and the wind to create limitless amounts of hydrogen to power the entire planet forever. If that somehow still proves not to be sufficient, civilization is learning how to grow the energy itself, through the use of a variety of plant life to produce ethanol and through the growing of algae to produce biodiesel. The most exciting of all is that the human race will not have to wait millenniums for this technology to develop, because all of these technologies are here now and will be as common as gasoline and electricity is today; certainly well before the twenty-first century is up.

Within fifty to one hundred years, and possibly much sooner, this clean, limitless energy will be abundant and available to everyone, everywhere on the planet. Exactly what does all that mean? For one, there will be no more pollution. The atmosphere, the oceans, the rivers, the lakes and the forests, will finally be able to begin the healing process, reversing 200 years of destruction. The poisonous gases in the atmosphere, the pollutants in the rivers and lakes, and the acid rain abuse of the forests will become a thing of the past.

Further, there will be no more unequal distribution of resources, thus creating a planet where most of the nations begin to enjoy similar standards of living. And finally, no more "oil wars." With resources distributed equally throughout the planet, equally available to everyone, there will be no need to guard our energy supply, as is presently needed in the Middle East. There will be no need to send young men and women to die, to protect American wasteful practices, in the name of protecting the U.S. way of life and the U.S. standard of living. Finally, those who died in the U.S. oil wars will be able to rest in peace.

This is not a futuristic, ultra-modern scenario. This is a fact. There is no doubt that this planet, once individual governments learn to cooperate, will provide more energy than the planet can use, and that this energy will be provided at a considerably lower cost than everyone pays today. The problem is how to get there. Will the U.S. government be capable of making the transformation painlessly, providing leadership for a smooth plan with a reasonable timetable for transformation, in which Americans continue to prosper while enjoying a high standard of living? Or is the U.S. government going to take a path where Americans will suffer a misery index our people never imagined; unaffordable energy prices, economic instability, recessions, depressions, unemployment, wars and a much lower standard of living, by the time the transformation is over? The journey to energy independence does not have to be miserable. The choice is up to the U.S. government and the American people.

So what are you waiting for? Pick up a phone and demand of your government representative to adopt the plan of "From Oil Hostage to Oil Freedom in a Generation."

Index

ANWR, 22, 84, 106, 197
ASPO, 32
atmosphere, 13
Bernard Cohen, 109
Bill Clinton, 85
Biodiesel, 121
Biofuel, 121
Biomass, 120, 121
Biomass Energy and Alcohol Fuels Act, 81
Coal, 11, 107, 147, 212
Conservation, 101, 178, 191, 196, 197, 224, 226
Crude oil, 19
Eisenhower, 78
Electric power, 144, 146
Electricity, 145, 158
Energy Information Administration, 8, 32, 163, 164
Energy Policy Act of 1992, 84
Energy Policy and Conservation Act, 79, 171, 177
Energy Security Act, 81
Ethanol, 3, 122, 215
Fission, 110
Fossil fuel, 148
Fuel cell, 138, 139
fuel economy standards, 172, 174, 180
Gasohol, 122
George W. Bush, 86
Geothermal energy, 127

Geothermal Energy Act, 81
Hover Dam, 124
Hydrogen, 131, 133, 134, 135, 136, 140, 142, 143, 202, 233
Hydrogen Fuel Initiative, 140
Hydropower, 123, 140, 147
Jimmy Carter, 53, 86, 95, 189
Mandatory Oil Import Quotas, 55
Middle East, 21, 23, 26, 28, 32, 41, 43, 45, 55, 56, 59, 65, 66, 71, 76, 83, 87, 90, 92, 106, 208, 217, 237
National Energy Strategy, 83
National Highway Traffic Safety Administration, 172, 174, 179
Natural gas, 106, 107, 189
Natural gushers, 18
Natural lift, 18
Niagara Falls, 124, 125
Nuclear energy, 110
nuclear power reactor, 112, 128, 135
OAPEC, 55

Ocean Thermal Energy Conversion Act, 81
Oil shale, 11, 27
OPEC, 43, 54, 55, 56, 57, 58, 59, 71, 72, 73, 77, 93, 95, 113, 169, 207, 208, 219, 220, 228, 229, 230
Partnership for a New Generation of Vehicles, 85
Petroleum, 11, 16
Plug-in hybrid electric vehicle, 159
Protectionism, 62
Ronald Reagan, 83, 85, 86
Simple distillation, 19
Solar Energy and Energy Conservation Act, 81
Tariffs, 64
The Corporate Average Fuel Economy Program, 171
United States Geological Survey, 32
Uranium, 109
Wind turbine, 114

Printed in the United States
82245LV00012B/142-153